A Concise Introduction to
Models and Methods for
Automated Planning

Synthesis Lectures on Artificial Intelligence and Machine Learning

Editor
Ronald J. Brachman, *Yahoo! Labs*
William W. Cohen, *Carnegie Mellon University*
Peter Stone, *University of Texas at Austin*

A Concise Introduction to Models and Methods for Automated Planning
Hector Geffner and Blai Bonet
2013

Essential Principles for Autonomous Robotics
Henry Hexmoor
2013

Case-Based Reasoning: A Concise Introduction
Beatriz López
2013

Answer Set Solving in Practice
Martin Gebser, Roland Kaminski, Benjamin Kaufmann, and Torsten Schaub
2012

Planning with Markov Decision Processes: An AI Perspective
Mausam and Andrey Kolobov
2012

Active Learning
Burr Settles
2012

Computational Aspects of Cooperative Game Theory
Georgios Chalkiadakis, Edith Elkind, and Michael Wooldridge
2011

Essentials of Game Theory: A Concise Multidisciplinary Introduction
Kevin Leyton-Brown and Yoav Shoham
2008

A Concise Introduction to Multiagent Systems and Distributed Artificial Intelligence
Nikos Vlassis
2007

Intelligent Autonomous Robotics: A Robot Soccer Case Study
Peter Stone
2007

A Concise Introduction to Models and Methods for Automated Planning
Hector Geffner and Blai Bonet

ISBN: 978-3-031-00436-0   paperback
ISBN: 978-3-031-01564-9   ebook

DOI 10.1007/978-3-031-01564-9

A Publication in the Springer series
SYNTHESIS LECTURES ON ARTIFICIAL INTELLIGENCE AND MACHINE LEARNING

Lecture #22
Series Editors: Ronald J. Brachman, *Yahoo! Labs*
                William W. Cohen, *Carnegie Mellon University*
                Peter Stone, *University of Texas at Austin*
Series ISSN
Synthesis Lectures on Artificial Intelligence and Machine Learning
Print 1939-4608   Electronic 1939-4616

A Concise Introduction to Models and Methods for Automated Planning

Hector Geffner
ICREA and Universitat Pompeu Fabra, Barcelona, Spain

Blai Bonet
Universidad Simón Bolívar, Caracas, Venezuela

SYNTHESIS LECTURES ON ARTIFICIAL INTELLIGENCE AND MACHINE LEARNING #22

ABSTRACT

Planning is the model-based approach to autonomous behavior where the agent behavior is derived automatically from a model of the actions, sensors, and goals. The main challenges in planning are computational as all models, whether featuring uncertainty and feedback or not, are intractable in the worst case when represented in compact form. In this book, we look at a variety of models used in AI planning, and at the methods that have been developed for solving them. The goal is to provide a modern and coherent view of planning that is precise, concise, and mostly self-contained, without being shallow. For this, we make no attempt at covering the whole variety of planning approaches, ideas, and applications, and focus on the essentials. The target audience of the book are students and researchers interested in autonomous behavior and planning from an AI, engineering, or cognitive science perspective.

KEYWORDS

planning, autonomous behavior, model-based control, plan generation and recognition, MDP and POMDP planning, planning with incomplete information and sensing, action selection, belief tracking, domain-independent problem solving

Contents

Preface

Planning is a central area in Artificial Intelligence concerned with the automated generation of behavior for achieving goals. Planning is also one of the oldest areas in AI with the General Problem Solver being the first automated planner and one of the first AI programs [Newell et al., 1959]. As other areas in AI, planning has changed a great deal in recent years, becoming more rigorous, more empirical, and more diverse. Planners are currently seen as automated solvers for precise classes of mathematical models represented in compact form, that range from those where the state of the environment is fully known and actions have deterministic effects, to those where the state of the environment is partially observable and actions have stochastic effects. In all cases, the derivation of the agent behavior from the model is computational intractable, and hence a central challenge in planning is scalability. Planning methods must exploit the structure of the given problems, and their performance is assessed empirically, often in the context of planning competitions that in recent years have played an important role in the area.

In this book, we look at a variety of models used in AI planning and at the methods that have been developed for solving them. The goal is to provide a modern and coherent view of planning that is precise, concise, and mostly self-contained, without being shallow. For this, we focus on the essentials and make no attempt at covering the whole variety of planning approaches, ideas, and applications. Moreover, our view of the essentials is not neutral, having chosen to emphasize the ideas that we find most basic in a model-based setting. A more comprehensive treatment of planning, circa 2004, can be found in the planning textbook by Ghallab et al. [2004]. Planning is also covered at length in the AI textbook by Russell and Norvig [2009].

The book is organized into eight chapters. Chapter 1 is about planning as the model-based approach to autonomous behavior in contrast to appproaches where behaviors are learned, evolved, or specified by hand. Chapters 2 and 3 are about the most basic model in planning, classical planning, where a goal must be reached from a fully known initial state by applying actions with deterministic effects. Classical planners can currently find solutions to problems over huge state spaces, yet many problems do not comply with these restrictions. The rest of the book addresses such problems in two ways: one is by automatically translating non-classical problems into classical ones; the other is by defining native planners for richer models. Chapter 4 focuses thus on reductions for dealing with soft goals, temporally extended goals, incomplete information, and a slightly different task: goal recognition. Chapter 5 is about planning with incomplete information and partial observability in a logical setting where uncertainty is represented by sets of states. Chapters 6 and 7 cover probabilistic planning where actions have stochastic effects, and the state is either fully or partially observable. In all cases, we distinguish between offline solution methods that derive the complete control offline, and online solution methods that derive the control as needed, by interleaving planning and execution, thinking and doing. Chapter 8 is about open problems.

We are grateful to many colleagues, co-authors, teachers, and students. Among our teachers, we would like to mention Judea Pearl, who was the Ph.D. advisor of both of us at different times, and always a role model as a person and as a scientist. Among our students, we thank in particular Hector

Palacios, Emil Keyder, Alex Albore, Miquel Ramírez, and Nir Lipovetzky, on whose work we have drawn for this book. The book is based on tutorials and courses on planning that one of us (Hector) has been giving over the last few years, more recently at the ICAPS Summer School (Thessaloniki, 2009; São Paulo, 2012; Perugia, 2013), the International Joint Conference on AI (IJCAI, Barcelona, 2011), La Sapienza, Università di Roma (2010), and the Universitat Pompeu Fabra (2012). We thank the students for the feedback and our colleagues for the invitations and their hospitality. Thanks also to Alan Fern who provided useful and encouraging feedback on a first draft of the book.

A book, even if it is a short one, is always a good excuse for remembering the loved ones.

A los chicos, caminante no hay camino, a Lito, la llama eterna, a Marina, mucho más que dos, a la familia toda; a la memoria del viejo, la vieja, la bobe, y los compañeros tan queridos − Hector

A Iker y Natalia, por toda su ayuda y amor, a la familia toda, por su apoyo. A la memoria de Josefina Gorgal Caamaño y la iaia Francisca Prat − Blai

Hector Geffner, Barcelona
Blai Bonet, Caracas
June 2013

CHAPTER 1

Planning and Autonomous Behavior

Planning is the model-based approach to autonomous behavior where the agent selects the action to do next using a model of how actions and sensors work, what is the current situation, and what is the goal to be achieved. In this chapter, we contrast programming, learning, and model-based approaches to autonomous behavior, and present some of the models in planning that will be considered in more detail in the following chapters. These models are all general in the sense that they are not bound to specific problems or domains. This generality is intimately tied to the notion of intelligence which requires the ability to deal with new problems. The price for generality is computational: planning over these models when represented in compact form is intractable in the worst case. A main challenge in planning is thus the automated exploitation of problem structure for scaling up to large and meaningful instances that cannot be handled by brute force methods.

1.1 AUTONOMOUS BEHAVIOR: HARDWIRED, LEARNED, AND MODEL-BASED

At the center of the problem of intelligent behavior is the *problem of selecting the action to do next*. In Artificial Intelligence (AI), three different approaches have been used to address this problem. In the *programming-based approach*, the controller that prescribes the action to do next is given by the programmer, usually in a suitable high-level language. In this approach, the problem is solved by the programmer in his head, and the solution is expressed as a program or as a collection of rules or behaviors. In the *learning-based approach*, the controller is not given by a programmer but is induced from experience as in reinforcement learning. Finally, in the *model-based approach*, the controller is not learned from experience but is derived automatically from a model of the actions, sensors, and goals. In all these approaches, the controller is the solution to the model.

The three approaches to the action selection problem are not orthogonal, and exhibit different virtues and limitations. Programming agents by hand, puts all the burden on the programmer that cannot anticipate all possible contingencies, and often results in systems that are brittle. Learning methods have the greatest promise and potential, but their flexibility is often the result of learning a model. Last, model-based methods require a model of the actions, sensors, and goals, and face the computational problem of solving the model—a problem that is computationally intractable even for the simplest models where information is complete and actions are deterministic.

The Wumpus game, shown in Figure 1.1 from the standard AI textbook [Russell and Norvig, 2009], is an example of a simple scenario where an agent must process information arriving from the sensors to decide what to do at each step. The agent, initially at the lower left corner, must obtain the gold while avoiding deadly pits and a killer wumpus. The locations of the gold, pits, and wumpus are

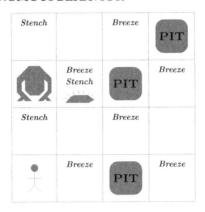

Figure 1.1: Autonomous Behavior in the Wumpus World: What to do next?

not known to the agent, but each emits a signal that can be perceived by the agent when in the same cell (gold) or in a contiguous cell (pits and wumpus). The agent control must specify the action to be done by the agent as a function of the observations gathered. The three basic approaches for obtaining such a controller are to write it by hand, to learn it from interactions with a Wumpus simulator, or to derive it from a model representing the initial situation, the actions, the sensors, and the goals.

While planning is often defined as the branch of AI concerned with the "synthesis of plans of action to achieve goals," planning is best conceived as *the model-based approach to action selection*—a view that defines more clearly the role of planning in intelligent autonomous systems. The distinction that the philosopher Daniel Dennett makes between "Darwinian," "Skinnerian," and "Popperian" creatures [Dennett, 1996], mirrors quite closely the distinction between hardwired (programmed) agents, agents that learn, and agents that use models respectively. The contrast between the first and the latter corresponds also to the distinction made in AI between reactive and deliberative systems, as long as deliberation is not reduced to logical reasoning. Indeed, as we will see, the inferences captured by model-based methods that scale up are not logical but heuristic, and follow from relaxations and approximations of the problem being solved.

PLANNING IS MODEL-BASED AUTONOMOUS BEHAVIOR

Model-based approaches to the action selection problem are made up of three parts: the *models* that express the dynamics, feedback, and goals of the agent; the *languages* that express these models in compact form; and the *algorithms* that use the representation of the models for generating the behavior.

A representation of the model for the Wumpus problem, for example, will feature *variables* for the locations of the agent, the gold, the wumpus, the pits, and a boolean variable for whether the agent is alive. The location variables can take 16 different values, corresponding with the cells in the 4×4 grid, except for the gold that can also be held by the agent and hence has 17 possible values.[1] A *state* for the problem is a valuation over these seven variables. The number of possible states is thus $16^5 \times 17 \times 2$, which is slightly more than 35 million. Initially, the agent is alive and knows its location

[1]If the number of pits and wumpus is not known a priori, an alternative representation would be needed where each cell in the grid would contain a wumpus, a pit, or neither.

but not the value of the pit and wumpus variables. The state of the system is thus *not fully observable*. The agent gets partial knowledge about the hidden variables through each of its three *sensors* that relate the true but hidden state of the world with observable tokens. The agent receives the observation token "stench" in the states where the wumpus is in one of the (at most) four cells adjacent to the agent, the token "breeze" in the states where a pit is adjacent to the agent, and the token "bright" when the gold and the agent are in the same cell. The actions available to the agent are to move to an adjacent cell, and to pick up the gold if known to be in the same cell. The actions change the state of the system in the expected way, affecting the location of the agent or the location of the gold. Yet the agent dies if it enters a cell with a wumpus or a pit, and a dead agent cannot execute any of the actions, and hence cannot achieve the goal of getting the gold.

In this problem, an intelligent agent should notice first that there is no wumpus or pit in cells $(1, 2)$ or $(2, 1)$ as there is no stench or breeze at the initial agent location $(1, 1)$. It is then safe to move either up or right. If it moves up, it'll sense a stench at $(1, 2)$ and conclude that the wumpus is at either $(1, 3)$ or $(2, 2)$. Likewise, since it senses no breeze, it can conclude that neither of these cells contains a pit. The only safe move is then to get back to $(1, 1)$ where it can move safely to $(2, 1)$. From the sensed breeze, it can conclude that there is a pit at $(3, 1)$ or $(2, 2)$, or one pit at each, and from sensing no stench, that there is no wumpus at either $(3, 1)$ or $(2, 2)$. At this point, it should conclude that cell $(2, 2)$ is safe as it cannot contain either a wumpus or a pit. It should then move up to $(2, 2)$, from which the process of visiting new cells that are safe is repeated until the gold is found.

Writing a program for solving any instance of the Wumpus domain, for any (solvable) initial situation and grid size, is interesting enough. Yet, the task in planning is quite different. We want a program that can take a representation of *any problem* exhibiting a certain mathematical structure, not limited to the Wumpus domain, and find a solution to it. A number of planning models will make these mathematical structures explicit. Other problems that have a number of features in common with the Wumpus domain include the familiar Battleship game or the popular PC game Minesweeper. These are all problems where a goal is to be achieved by acting and sensing in a world where the state of the system, that may change or not, is partially observable.

While a program that has been designed to play the Wumpus can be deemed as intelligent, a program that can play the Wumpus without having been designed specifically for it will be intelligent in a much broader sense. The first contains the recipes for playing the Wumpus; the latter contains "recipes" for playing an infinite collection of domains, known or unknown to the programmer, that share a general mathematical structure. The formulation of these mathematical structures and the general "recipes" for solving them is what planning is about.

1.2 PLANNING MODELS AND LANGUAGES

A wide range of models used in planning can be understood as variations of a *basic state model* featuring:

- a finite and discrete state space S,

- a *known initial state* $s_0 \in S$,

- a non-empty set $S_G \subseteq S$ of goal states,

- actions $A(s) \subseteq A$ applicable in each state $s \in S$,

- a *deterministic* state transition function $f(a, s)$ such that $s' = f(a, s)$ stands for the state resulting of applying action a in s, $a \in A(s)$, and

- *positive action costs* $c(a, s)$.

This is the model underlying *classical planning* where it is normally assumed that action costs $c(a, s)$ do not depend on the state, and hence $c(a, s) = c(a)$. A solution or *plan* in this model is a sequence of applicable actions that map the initial state into a goal state. More precisely, a plan $\pi = a_0, \ldots, a_{n-1}$ must generate a state sequence $s_0, \ldots, s_n$ such that $a_i \in A(s_i)$, $s_{i+1} = f(a_i, s_i)$, and $s_n \in S_G$, for $i = 0, \ldots, n - 1$. The cost of the plan is the sum of the action costs $c(a_i, s_i)$, and a plan is optimal if it has minimum cost over all plans.

Classical planners accept a compact description of models of this form in *languages* featuring *variables*, where the states are the possible valuations of the variables. A classical plan $\pi = a_0, \ldots, a_n$ represents an *open-loop controller* where the action to be done at time step i depends just on the step index i. The solution of models that accommodate uncertainty and feedback, produce *closed-loop controllers* where the action to be done at step i depends on the actions and observations collected up to that point. These models can be obtained by relaxing the assumptions in the model above displayed in italics.

The model for *partially observable planning*, also called *planning with sensing* or *contingent planning*, is a variation of the classical model that features both *uncertainty* and *feedback*—namely, uncertainty about the initial and next possible state, and partial information about the current state of the system. Mathematically such a model can be expressed in terms of the following ingredients:

- a finite and discrete state space S,

- a non-empty set S_0 of *possible* initial states, $S_0 \subseteq S$,

- a non-empty set $S_G \subseteq S$ of goal states,

- a set of actions $A(s) \subseteq A$ applicable in each state $s \in S$,

- a *non-deterministic* state transition function $F(a, s)$ for $s \in S$ and $a \in A(s)$, where $F(a, s)$ is non-empty and $s'' \in F(a, s)$ stands for the possible successor states of state s after action a is done, $a \in A(s)$,

- a set of observation tokens O,

- a sensor model $O(s, a) \subseteq O$, where $o \in O(s, a)$ means that token o may be observed in the (possibly hidden) state s if a was the last action done, and

- positive *action costs* $c(a, s)$.

In the model for the Wumpus problem, the state space S is given by the set of possible valuations over the problem variables, S_0 is the set of states where the agent is initially alive and at location $(1, 1)$, S_G is the set of states where the agent is holding the gold, and A stands for the actions of moving and picking up the gold, provided that the agent can't leave the grid and can't pick the gold if not in the same cell. Likewise, the state transitions $F(a, s)$ associated with these actions is deterministic, meaning that $F(a, s)$ contains a single state s' so that $|F(a, s)| = 1$. The same is true for the sensor

Planning Problem ⟶ $\boxed{Planner}$ ⟶ *Controller* ⟷ $\boxed{Environment}$

Figure 1.2: A planner takes a compact representation of a planning problem over a certain class of models (classical, conformant, contingent, MDP, POMDP) and automatically produces a controller. For fully and partially observable models, the controller is closed-loop, meaning that the action selected depends on the observations gathered. For non-observable models like classical and conformant planning, the controller is open-loop, meaning that it is a fixed action sequence.

model $O(s, a)$, which does not depend on a but just on the hidden state s. Namely, O contains nine observation tokens o, corresponding to the possible combinations of the three booleans stench, breeze, and bright, so that if s is a state where the agent is next to a pit and a wumpus but not in the same cell as the gold, then $o \in O(s, a)$ iff o represents the combination where stench and breeze are true, and bright is false. The action costs for the problem, $c(a, s)$, can be all assumed to be 1, and in addition, no action can be done by the agent when he is not alive.

A partially observable planner is a program that accepts compact descriptions of instances of the model above, like the one for the Wumpus, and automatically outputs the control (Figure 1.2). As we will see, planners come in two forms: *offline* and *online*. In the first case, the behavior specifies the agent response to each possible situation that may result; in the second case, the behavior just specifies the action to be done in the current situation. These types of control, unlike the control that results in classical planning, are closed-loop: the actions selected usually depend on the observation tokens received.

Offline solutions of partially observable problems are not fixed action sequences as in classical planning, as observations need to be taken into account for selecting actions. Mathematically, thus, these solutions are functions mapping the stream of past actions and observations into actions, or more conveniently, functions mapping *belief states* into actions. The belief state that results after a given stream of actions and observations represents the set of states that are deemed possible at that point, and due to the Markovian state-transition dynamics, it summarizes all the information about the past that is relevant for selecting the action to do next. Moreover, since the initial belief state b_0 is given, corresponding to the set of possible initial states S_0, a solution function π, called usually the *control policy*, does not need to be defined over all possible beliefs, but just over the beliefs that can be produced from the actions determined by the policy π from the initial belief state b_0 and the observations that may result. Such *partial* policies π can be represented by a directed graph rooted at b_0, where nodes stand for belief states, edges stand for actions a_i or observations o_i, and the branches in the graph from b_0, stand for the stream of actions and observations $a_0, o_0, a_1, o_1, \ldots$, called *executions*, that are possible. The policy solves the problem when all these possible executions end up in belief states where the goal is true.[2]

The models above are said to be *logical* as they only encode and keep track of what is possible or not. In *probabilistic* models, on the other hand, each possibility is weighted by a probability measure. A probabilistic version of the partially observable model above can be obtained by replacing the set of possible initial states S_0, the set of possible successor states $F(a, s)$, and the set of possible observation tokens $O(s, a)$, by probability distributions: a prior $P(s)$ on the states $s \in S_0$ that are initially possible,

---

[2]We will make this all formal and precise in Chapter 5.

transition probabilities $P_a(s'|s)$ for encoding the likelihood that s' is the state that follows s after a, and observation probabilities $P_a(o|s)$ for encoding the likelihood that o is the token that results in the state s when a is the last action done.

The model that results from changing the *sets* S_0, $F(a, s)$, and $O(s, a)$ in the partially observable model, by the probability distributions $P(s)$, $P_a(s'|s)$, and $P_a(o|s)$, is known as a *Partially Observable Markov Decision Process* or POMDP [Kaelbling et al., 1998]. The advantages of representing uncertainty by probabilities rather than sets is that one can then talk about the *expected cost* of a solution as opposed to the cost of the solution in the *worst case*. Indeed, there are many meaningful problems that have infinite cost in the worst case but perfectly well-defined expected costs. These include, for example, the problem of preparing an omelette with an infinite collection of eggs that may be good or bad with non-zero probabilities, but that can be picked up and sensed one at a time. Indeed, while the scope of probabilistic models is larger than the scope of logical models, we will consider both, as the latter are simpler, and the computational ideas are not all that different.

A *fully observable* model is a partially observable model where the state of the system is fully observable, i.e., where $O = S$ and $O(s, a) = \{s\}$. In the logical setting such models are known as *Fully Observable Non-Deterministic* models, abbreviated FOND. In the probabilistic setting, they are known as *Fully Observable Markov Decision Processes* or MDPs [Bertsekas, 1995].

Finally, an *unobservable* model is a partially observable model where no relevant information about the state of the system is available. This can be expressed through a sensor model O containing a single dummy token o that is "observed" in all states, i.e., $O(s, a) = O(s', a) = \{o\}$ for all s, s', and a. In planning, such models are known as *conformant*, and they are defined exactly like partially observable problems but with *no sensor* model. Since there are no (true) observations, the solution form of conformant planning problems is like the solution form of classical planning problems: a fixed action sequence. The difference between classical and conformant plans, however, is that the former must achieve the goal for the given initial state and unique state-transitions, while the latter must achieve the goal in spite of the uncertainty in the initial situation and dynamics, for *any possible initial state* and *any state transition that is possible*. As we will see, conformant problems make up an interesting stepping stone in the way from classical to partially observable planning.

In the book, we will consider each of these models in turn, some useful special cases, and some variations. This variety of models is the result of several orthogonal dimensions: uncertainty in the initial system state (fully known or not), uncertainty in the system dynamics (deterministic or not), the type of feedback (full, partial or no state feedback), and whether uncertainty is represented by sets of states or probability distributions.

1.3   GENERALITY, COMPLEXITY, AND SCALABILITY

Classical planning, the simplest form of planning where actions have deterministic effects and the initial state is fully known, can be easily cast as a path-finding problem over a directed graph where the nodes are the states, the initial node and target nodes are the initial and goal states, and a directed edge between two nodes denotes the existence of an action that maps one state into the other. Classical planning problems can thus be solved in theory by standard path-finding algorithms such as Dijkstra's that run in time that is polynomial in the number of nodes in the graph [Cormen et al., 2009, Dijkstra, 1959]. Yet in planning, this is not good enough as the nodes in the graph stand for the problem states, whose number is exponential in the number of problem variables. If these variables have at least two

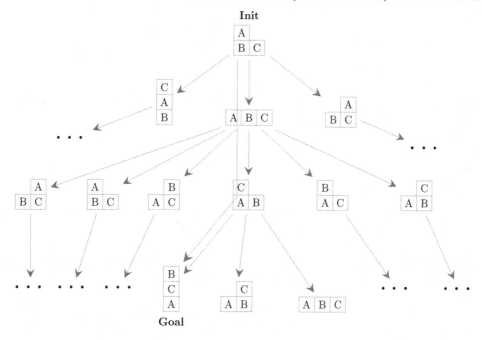

Figure 1.3: The graph corresponding to a simple planning problem involving three blocks with initial and goal situations as shown. The actions allow to move a clear block on top of another clear block or to the table. The size of the complete graph for this domain is exponential in the number of blocks. A plan for the problem is shown by the path in red.

possible values, the number of nodes in the graph to search can be in the order of 2^n, where n is the number of variables. In particular, if the problem involves 30 variables, this means $1,073,741,824$ nodes, and if the problem involves 100 variables, it means more than 10^{30} nodes. In order to get a concrete idea of what exponential growth means, if it takes one second to generate 10^7 nodes (a realistic estimate given current technology), it would take more than 10^{23} seconds to generate 10^{30} nodes. This is however almost one million times the estimated age of the universe.[3]

A more vivid illustration of the complexity inherent to the planning problem can be obtained by considering a well known domain in AI: the Blocks World. Figure 1.3 shows an instance of this domain where blocks A, B, and C, initially arranged so that A is on B, and B and C are on the table, must be rearranged so that B is on C, and C is on A. The actions allow to move a clear block (a block with no block on top) on top of another clear block or on the table. The problem can be easily expressed as a classical planning problem where the variables are the block locations: blocks can be on the table or on top of another block. The figure shows the graph associated to the problem whose solution is a path connecting the node representing the initial situation with a node representing a goal situation. The number of states in a Blocks World problem with n blocks is exponential in n, as the states include all the $n!$ possible towers of n blocks plus additional combinations of lower towers. Thus, a planner

---

[3]The age of the universe is estimated at 13.7×10^9 years approximately. Visiting 2^{100} nodes at 10^7 nodes a second would take in the order of 10^{15} years, as $2^{100}/(10^7 * 60 * 60 * 24 * 365) = 4.01969368 \times 10^{15}$.

able to solve arbitrary Blocks World instances should be able to search for paths over huge graphs. This is a crisp computational challenge that is very different from writing a *domain-specific* Blocks World solver—namely, a program for solving any instance of this specific domain. Such a program could follow a domain-specific strategy, like placing all misplaced blocks on the table first, in order, from top to bottom, then moving these blocks to their destination in order again, this time, from the bottom up. This program will solve any instance of the Blocks World but will be completely useless in other domains. The challenge in planning is to achieve both *generality* and *scalability*. That is, a classical planner must accept a description of *any problem* in terms of a set of variables whose initial values are known, a set of actions that change the values of these variables deterministically, and a set of goals defined over these variables. The planner is domain-general or domain-independent in the sense that it does not know what the variables, actions, and domain stand for, and for any such description it must decide effectively which actions to do in order to achieve the goals.

For classical planning, as for the other planning models that we will consider, the general problem of coming up with a plan is NP-hard [Bylander, 1994, Littman et al., 1998]. In Computer Science, an NP-hard problem (non-deterministic polynomial-time hard) is a problem that is at least as hard as any NP-complete problem; these are problems that can be solved in polynomial time by a non-deterministic Turing Machine but which are widely believed not to admit polynomial-time solutions on deterministic machines [Sipser, 2006]. The complexity of planning and related models has been used as evidence for contesting the possibility of general planning and reasoning abilities in humans or machines [Tooby and Cosmides, 1992]. The complexity of planning, however, just implies that no planner can efficiently solve every problem from every domain, not that a planner cannot solve an infinite collection of problems from seen and unseen domains, and hence be useful to an acting agent. This is indeed the way modern AI planners are empirically evaluated and ranked in the AI planning competitions, where they are tried over domains that the planners" authors have never seen. Thus, far from representing an insurmountable obstacle, the twin requirements of generality and scalability have been addressed head on in AI planning research, and have resulted in simple but powerful computational principles that make domain-general planning feasible. The computational challenge aimed at achieving both scalability and generality over a broad class of intractable models, has actually come to characterize a lot of the research work in AI, that has increasingly focused on the development of effective algorithms or *solvers* for a wide range of tasks and models (Figure 1.4); tasks and models that include SAT and SAT-variants like Weighted-Max SAT and Weighted Model Counting, Bayesian Networks, Constraint Satisfaction, Answer Set Programming, General Game Playing, and Classical, MDP, and POMDP Planning. This is all work driven by theory and experiments, with regularly held competitions used to provide focus, to assess progress, and to sort out the ideas that work best empirically [Geffner, 2013a].

1.4   EXAMPLES

We consider next a simple navigation scenario to illustrate how different types of planning problems call for different planning models and different solution forms. The general scenario is shown in Figure 1.5 where the agent marked as A has to reach the goal marked as G. The four actions available let the agent move one unit in each one of the four cardinal directions, as long as there is no wall. Actions that lead the agent to a wall have no effect. The question is how should the agent select the actions for achieving the goal with certainty under different knowledge and sensing conditions. In all

$$Model\ Instance \longrightarrow \boxed{Solver} \longrightarrow Solution$$

Figure 1.4: Models and Solvers: Research work in AI has increasingly focused on the formulation and development of solvers for a wide range of models. A solver takes the representation of a model instance as input, and automatically computes its solution in the output. Some of the models considered are SAT, Bayesian Networks, General Games Playing, and Classical, MDP, and POMDP Planning. All of these models are intractable when represented in compact form. The main challenge is scaling up.

cases, we assume that the agent knows the map, including where the walls and the goals are. In the simplest case, the actions are assumed to be deterministic and the initial agent location known. In this case, the agent faces a classical planning problem whose solution is a path in the grid joining the initial agent location and the goal. On the other hand, if the actions have effects that can only be predicted probabilistically but the state of the problem—the agent location—is always observable, the problem becomes an MDP planning problem. The solution to this problem is no longer an action sequence, that cannot guarantee that the goal will be achieved with certainty, but a policy assigning one of the four possible actions to each one of the states. The number of steps to reach the goal can no longer be determined with certainty but there is then an expected number of steps to reach the goal that can be determined, as the policy and the action model induce a probability distribution over all the possible paths in the grid. Policies that ensure that the goal is eventually achieved with certainty are called *proper policies*. Interestingly, we will see that the exact transition probabilities are not relevant for defining or computing proper policies; all we need to know for this are which state transitions are possible (probability different than zero) and which ones are not (probability equal to zero). The problem variation in which the actions have stochastic effects but the location of the agent cannot be fully observed is a POMDP planning problem, whose general solution is neither a fixed action sequence, that ignores the observations, nor a policy prescribing the action to do in each state, that assumes that the state is observable. It is rather a policy that maps *belief states* into actions. In the POMDP setting, a belief state is a probability distribution over the states that are deemed possible. These probability distributions summarize all the information about the past that is relevant for selecting the action to do next. The initial belief state has to be given, and the current belief state is determined by the actions done, the observations gathered, and the information in the model. On the other hand, if uncertainty about the initial situation, the system dynamics and the feedback are represented by *sets of states* as opposed to *probability distributions over states*, we obtain a *partially observable planning problem* which is the logical counterpart of POMDPs. In such problems, the number of possible belief states (sets of states deemed possible) is finite, although exponential in the number of states, and hence doubly exponential in the number of problem variables. Finally, if the agent must reach the goal with certainty but there is uncertainty about the initial state or about the next state dynamics, and there is *no feedback* of any type, the problem that the agent faces is a *conformant problem*. For the problem shown in the figure, if the actions are deterministic and the agent knows that it is initially somewhere on the room on the left, a conformant solution can be obtained as follows: the agent moves up five times, until it knows with certainty that it is somewhere on the top row, then it moves right three times until it knows with certainty that it is exactly at the top right corner of the left room. With all uncertainty gone, the agent can then find a path to the goal from that corner.

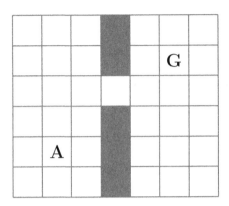

Figure 1.5: Variations on a planning problem: Agent, marked as A, must reach the goal marked G, by moving one cell at a time under different knowledge and sensing conditions.

A completely different planning example is shown in Figure 1.6 for a problem inspired on the use of deictic representations [Ballard et al., 1997, Chapman, 1989], where a visual-marker or "eye" (the circle on the lower left) must be placed on top of a green block by moving it one cell at a time. The location of the green block is not known initially, and the observations are just whether the cell with the mark contains a green block (G), a non-green block (B), or neither (C), and whether such cell is at the level of the table (T) or not (–). The problem is a partially observable planning problem and the solution to it can be expressed by means of control policies mapping beliefs into actions. An alternative way to represent solutions to these types of problems is by means of *finite-state controllers*, such as the one shown on the right of Figure 1.6. This finite-state controller has two internal states, the initial controller state q_0, and a second controller state q_1. An arrow $q \overset{o/a}{\to} q'$ in the controller indicates that, when obtaining the observation o in the controller state q, the action a should be performed, moving to the controller state q', that may be equal to q or not, and where the same action selection mechanism is applied. The reader can verify that the finite-state controller searches for a tower with a green block from left to right, going all the way up to the top block in each tower, then going all the way down to the table, and iterating in this manner until the visual marker appears on top of a block that is green. Finite-state controllers provide a very compact and convenient representation of the actions to be selected by an autonomous system, and for this reason they are commonly used in practice for controlling robots or non-playing characters in video games [Buckland, 2004, Mataric, 2007, Murphy, 2000]. While these controllers are normally written by hand, we will show later that they they can be obtained automatically using planners. Indeed, the controller shown in the figure has been derived in this way using a *classical planner* over a suitable transformation of the partially observable problem shown on the left [Bonet et al., 2009]. It is actually quite remarkable that the finite-state controller that has been obtained in this manner is not only good for solving the original problem on the left, but also an infinite number of variations of it. It can actually be shown that the controller will successfully solve any modification in the problem resulting from *changes in either the dimensions of the grid, the number of blocks or their configuration*. Thus, in spite of appearances, the power

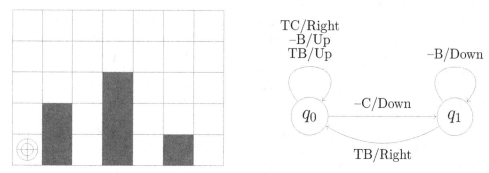

Figure 1.6: *Left:* Problem where a visual-marker (mark on the lower left cell) must be placed on top of a green block by just observing what is on the marked cell. *Right:* Finite-state controller obtained with a classical planner from suitable translation. The controller solves the problem and any variation of the problem that results from changes in the number or configuration of blocks.

of classical planners shouldn't be underestimated. Often we will be able to solve non-classical planning problems using classical planners by means of feasible, well-defined transformations.

1.5   GENERALIZED PLANNING: PLANS VS. GENERAL STRATEGIES

The visual-marker problem illustrates two differences that are crucial in planning. One is the difference between a *solution to a problem instance* and *a solution to a family of instances*. The second is the difference between *expressing the solution* to a problem and *finding a solution* to the problem in the first place. For example, a solution to a particular Blocks World instance, with blocks A, B, and C, on the table, that must be stacked in order with C on top, may be the action sequence *pick(B)*, *stack(B,A)*, *pick(C)*, *stack(C,B)*. On the other hand, the *general strategy* of putting all blocks on the table, in order from top to bottom, followed by putting all blocks on their destination in order from the bottom to the top, works for all Blocks World instances, regardless of the number and names of the blocks. Planning in AI, and in particular what is called domain-independent planning, has been mostly focused on models and methods *for expressing and solving single planning instances*. On the other hand, the work in planning driven by applications over certain specific domains, has usually focused on languages like *Hierarchical Task Networks* [Erol et al., 1994] *for expressing by hand the strategies for solving any problem in the given domain*. The problem of computing general domain strategies has not been tackled by automated methods, because the problem appears to be too hard in general. Yet, ideally, this is where we would like to get, at least on domains that admit compact general solutions [Srivastava et al., 2011a]. In a recent formulation, a form of *generalized planning* of this type has been shown to be EXPSPACE-Complete [Hu and de Giacomo, 2011]. In this formulation, all instances are assumed to share the same set of actions and observations, and a general solution is a function mapping streams of observations into actions. In some cases, such functions can be conveniently expressed as policies that map suitable combination of observables, called *features*, into actions. The crucial question is how to get such features and policies effectively, in particular over domains that admit compact solutions over the right features. An early approach that does this in the blocks world constructs a pool of possible features from the primitive domain predicates and a simple grammar (a description logic), and then looks for compact rule-based policies over some of such features using supervised learning algorithms [Fern et al., 2003, Martin and Geffner, 2000]. Another approach, used for playing a challenging game in real-time, represents the policies that map observables into actions by means of neural networks whose topology and weights are found by a form of evolutionary search [Stanley et al., 2005]. Both of these approaches are aimed at general domain policies that map state features into actions, which are not tied to specific instances.

In a famous exchange during the 80s about "universal plans," Matthew Ginsberg attacked the value of the idea of general plans on computational grounds [Ginsberg, 1989]. A universal plan is a strategy for solving not just one planning instance, but many instances, and more specifically, all those instances that can be obtained by just changing the initial state of the problem [Schoppers, 1987]. While the solutions to such generalized planning problems can be expressed as policies mapping states into actions, as for MDPs, Ginsberg argued that the size of such universal plans would be often just too large; exponential at least for families of problems that are NP-hard to solve. In order to illustrate this point, Ginsberg conjectured that no compact universal plan could be defined for a specific problem that he called the fruitcake problem, where blocks were to be placed on a tower to spell the word "fruitcake." The challenge, however, was answered by David Chapman who came up with an elegant reactive architecture, basically a circuit in line with his "Pengi" system [Agre and Chapman, 1987], that in polynomial time solved the problem [Chapman, 1989]. Chapman went on to claim that *Blockhead*, his system, solved the fruitcake problem, and solved it easily, with no search or planning, thus raising doubts not only about the value of "universal plans" but also of the need to plan itself. This was an interesting debate, and if we bring it here it is because it relates to the two key distinctions mentioned at the beginning of this section: one between a *solution to a problem instance* and *a solution to a family of instances*, the other between *expressing a solution* and *finding a solution*. From this perspective, Chapman is right that general strategies for solving many classes of interesting problems can often be encoded through compact representations, like Pengi-like circuits, yet the challenge is in coming up with such strategies automatically. Without this ability, it cannot be said that Blockhead is solving the problem—it's Chapman. Blockhead simply executes the policy crafted by Chapman that is not general and doesn't apply to other domains.

1.6   HISTORY

The first AI planner and one of the first AI programs was introduced by Newell and Simon in the 50s [Newell and Simon, 1961, Newell et al., 1959]. This program called GPS, for General Problem Solver, introduced a technique called means-ends analysis where differences between the current state and the goal were identified and mapped into operators that decreased those differences. The STRIPS system [Fikes and Nilsson, 1971] combined means-ends analysis with a convenient declarative action language. Since then, the idea of means-ends analysis has been refined and extended in many ways, in the formulation of planning algorithms that are *sound* (only produce plans), *complete* (produce a plan if one exists), and *effective* (scale up to large problems). By the early 90s, the state-of-the-art planner was UCPOP [Penberthy and Weld, 1992], an implementation of an elegant planning method known as partial-order planning where plans are not searched either forward from the starting state or backward from the goal, but are constructed from a decomposition scheme in which joint goals are decomposed into subgoals that create as further subgoals the preconditions of the actions used to establish them

[McAllester and Rosenblitt, 1991, Sacerdoti, 1975, Tate, 1977]. The actions that are incorporated into the plan are partially ordered as needed in order to resolve possible conflicts among them. Partial-order planning algorithms are sound and complete, but do not scale up well, as there are too many choices to make and too little guidance on how to make those choices; yet see [Nguyen and Kambhampati, 2001, Vidal and Geffner, 2006].

The situation in planning changed drastically in the mid 90s with the introduction of Graphplan [Blum and Furst, 1995], an algorithm that appeared to have little in common with previous approaches but scaled up much better. Graphplan builds a *plan graph* in polynomial time reasoning forward from the initial state, which is then searched backward from the goal to find a plan. It was shown later that the reason Graphplan scaled up well was due to a powerful admissible heuristic implicit in the plan graph [Haslum and Geffner, 2000]. The success of Graphplan prompted other approaches. In the SAT approach [Kautz and Selman, 1996], the planning problem for a fixed planning horizon is converted into a general *satisfiability* problem expressed as a set of clauses (a formula in Conjunctive Normal Form or CNF) that is fed into state-of-the-art SAT solvers, which currently manage to solve huge SAT instances even though the SAT problem is NP-complete.

Currently, the formulation of classical planning that appears to scale up best is based on heuristic search, with heuristic values derived from the delete-relaxation [Bonet et al., 1997, McDermott, 1996]. In addition, state-of-the-art classical planners use information about the actions that are most "helpful" in a state [Hoffmann and Nebel, 2001], and implicit subgoals of the problem, called landmarks, that are also extracted automatically from the problem with methods similar to those used for deriving heuristics [Hoffmann et al., 2004, Richter and Westphal, 2010].

Since the 90s, increasing attention has been placed on planning over non-classical models such as MDPs and POMDPs where action effects are not fully predictable, and the state of the system is fully or partially observable [Dean et al., 1993, Kaelbling et al., 1998]. We will consider all these variations and others in the rest of the book. We will have less to say about *Hierarchical Task Planning* or HTN planning, which, while widely used in practice, is focused on the representation of *general strategies* for solving problems rather than in representing and solving the problems themselves. For a comprehensive planning textbook, see Ghallab et al. [2004], while for a modern AI textbook covering planning at length, see Russell and Norvig [2009].

CHAPTER 2

Classical Planning: Full Information and Deterministic Actions

In classical planning, the task is to drive a system from a given initial state into a goal state by applying actions whose effects are deterministic and known. Classical planning can be formulated as a path-finding problem over a directed graph whose nodes represent the states of the system or enviroment, and whose edges capture the state transitions that the actions make possible. The computational challenge in classical planning results from the number of states, and hence the size of the graph, which are exponential in the number of problem variables. State-of-the-art methods in classical planning search for paths in such graphs by directing the search toward the goal using *heuristic functions* that are automatically derived from the problem. The heuristic functions map each state into an *estimate* of the distance or cost from the state to the goal, and provide the search for the goal with a sense of direction. In this chapter, we look at the model and languages for classical planning, and at the heuristic search techniques that have been developed for solving it. Variations and extensions of these methods, as well as alternative methods, will be considered in the next chapter.

2.1 CLASSICAL PLANNING MODEL

Classical planning is concerned with the selection of actions in environments that are *deterministic* and whose initial state is *fully known*. The model underlying classical planning can be described as the state model $\mathcal{S} = \langle S, s_0, S_G, A, f, c \rangle$ where

- S is a finite and discrete set of states,

- $s_0 \in S$ is the *known initial state,*

- $S_G \subseteq S$ is the non-empty set of goal states,

- $A(s) \subseteq A$ represents the set of actions in A that are applicable in each state $s \in S$,

- $f(a, s)$ is the *deterministic transition function* where $s' = f(a, s)$ is the state that follows s after doing action $a \in A(s)$, and

- $c(a, s)$ is a *positive cost* for doing action a in the state s.

A solution or *plan* in this model is a sequence of applicable actions $a_0, \ldots, a_n$ that generates a state sequence $s_0, s_1, \ldots, s_{n+1}$ where s_{n+1} is a goal state. More precisely, the action a_i is applicable in the state s_i if $a_i \in A(s_i)$, the state s_{i+1} follows the state s_i if $s_{i+1} = f(a_i, s_i)$, and s_{n+1} is a goal state if $s_{n+1} \in S_G$. The cost of the plan is the sum of the action costs $c(a_i, s_i)$, $i = 0, \ldots, n$. A plan is optimal if it has minimum cost, and the cost of the model is the cost of an optimal plan. A common cost structure arises when all action costs $c(a, s)$ are equal to 1. Then the cost of the plan is given by its length, and the optimal plans are the shortest ones.

Domain-independent classical planners accept a compact description of the above models, and automatically produce a plan. The problem is intractable in the worst case [Bylander, 1994], yet currently large classical problems can be solved very quickly. Optimal planners produce optimal plans, while satisficing planners are aimed at producing good plans which are not necessarily optimal. Computing optimal plans is harder than computing plans, as the former involves a universal claim about the space of all plans—namely, that they all have a cost that is not smaller than the cost of the plan found.

2.2   CLASSICAL PLANNING AS PATH FINDING

There is a direct correspondence between classical planning models and directed graphs, and between classical plans and certain paths over these graphs. Recall that directed graphs are structures $G = (V, E)$ made up of a set of nodes V and a set of directed edges represented by ordered pairs of nodes (n, n'), $n, n' \in V$. Directed graphs or digraphs are a fundamental model in Computer Science, and there is a wide range of algorithms for solving tasks over graphs [Cormen et al., 2009]. A weighted digraph is a digraph $G = (V, E, w)$ where every edge (n, n') in E comes with a real weight $w(n, n')$. One of the basic tasks over digraphs is finding a (directed) path from a given source node n_0 to a node n in a target set, and similarly, one of the basic tasks over weighted digraphs is finding one such path with minimum cost.

A classical planning model $\mathcal{S} = \langle S, s_0, S_G, A, f, c \rangle$ defines a weighted directed graph $G = (V, E, w)$ where the nodes in V represent the states in S, the edges (s, s') in E represent the presence of an action a in $A(s)$ such that s' is the state that follows a in s, and the weight $w(s, s')$ is the minimum cost over such actions in s. It follows from this correspondence that an action sequence $\pi = a_0, \ldots, a_m$ is a plan for the state model $\mathcal{S}$ iff π generates a sequence of states $s_0, \ldots, s_{m+1}$ that represents a directed path in the weighted digraph $G = (V, E, w)$. Clearly, if one such path can be found in the graph, the corresponding plan can be obtained by retrieving a_i as the action that makes the transition (s_i, s_{i+1}) possible with least cost. Thus, in principle any *path-finding algorithm* over weighted directed graphs can be used for finding plans for the classical planning model. We look at some of these algorithms below. All these algorithms are incremental in the sense that none of them requires the graph to be explicit in memory at the beginning; rather they all explicate an implicit graph incrementally as the search to the goal proceeds. This is crucial in planning where the size of the graphs is exponential in the number of problem variables.

2.3   SEARCH ALGORITHMS: BLIND AND HEURISTIC

Path-finding algorithms come in two main varieties: those in which the goal plays an active role in the search, and those in which the goal sits passively until encountered. The standard way in which goals can bias the search is by means of *heuristic functions*; these are functions $h(s)$ that provide a quick-and-dirty estimate of the cost to reach the goal from the state s, making the search goal-directed. Algorithms that use these heuristics are called *heuristic search algorithms*; those in which the goals

---

SOLVE(*Nodes*)
   while *Nodes* $\neq \emptyset$ **do**
      Let $n :=$ SELECT-NODE(*Nodes*)
      Let *Rest* $:=$ *Nodes* $\setminus \{n\}$
      if n is a goal node **then**
         return EXTRACT-SOLUTION(n)
      else
         Let *Children* $:=$ EXPAND-NODE(n)
         Set *Nodes* $:=$ ADD-NODES(*Children*, *Rest*)
      end if
   end while
   return Unsolvable

Figure 2.1: General search schema invoked with *Nodes* containing the source node only. A number of familiar search algorithms are obtained by suitable choices of the SELECT-NODE and ADD-NODE functions.

play no active role during the search are called *brute force* or *blind search* algorithms [Edelkamp and Schrödl, 2012, Pearl, 1983]. The latter include algorithms like *Depth-First Search*, *Breadth-First Search*, and *Uniform Cost Search*, also called Dijkstra's algorithm [Cormen et al., 2009]. The former include *Best-First Search*, *A**, and *Hill Climbing*. The algorithms search for paths in different ways, and have different properties concerning completeness, optimality, and time and memory complexity. We will make a quick overview of them before reviewing some useful variants.

The algorithms can all be understood as particular instances of the general schema shown in Figure 2.1, where a search frontier called *Nodes*, initialized with the root node of the graph, shifts incrementally as the graph is searched. In each iteration, two steps are performed: a selected node is removed from the search frontier, and if the selected node is not a goal node, its children are added to the search frontier, else the search terminates and the path to the last node selected is returned. The nodes in the search represent the states of the problem and contain in addition bookkeeping information like a pointer to the parent node, and the weight of the best path to the node so far, called the *accumulated cost* and denoted by the expression $g(n)$ where n is the node. The various algorithms arise from the representation of the search frontier *Nodes*, the way nodes are selected from this frontier, and the way the children of these nodes are added to the search frontier.

Depth-First Search is the algorithm that results from implementing the search frontier *Nodes* as a STACK: the node that is selected is the top node in the stack, and the children nodes are added to the top of the stack. It can be shown that if the graph is acyclic, the nodes in *Nodes* will be selected indeed in depth-first order. Similarly, Breadth-First Search is the algorithm that results from implementing the search frontier *Nodes* as a QUEUE: nodes are selected from one end, and their children are added to the other end. It can be shown then that the nodes in *Nodes* will be selected depth-last, and more precisely, shallowest-first, which is the characteristic of Breadth-First Search. Finally, Best-First Search is the algorithm that results when the search frontier is set up as a PRIORITY-QUEUE, so that the nodes selected are the ones that minimize a given evaluation function $f(n)$. Best-First Search reduces to the well-known A* algorithm [Hart et al., 1968] when the evaluation function is defined as the sum

$f(n) = g(n) + h(n)$ of the accumulated cost $g(n)$ and the heuristic estimate of the cost-to-go $h(n)$. Other known variations of Best-First Search are Greedy Best-First where $f(n) = h(n)$, and WA* where $f(n) = g(n) + Wh(n)$ and W is a constant larger than 1. Finally, Uniform-Cost Search or Dijkstra's algorithm corresponds basically to a Best-First Search with evaluation function $f(n) = g(n)$, or alternatively to the A* algorithm with the heuristic function $h(n)$ set to 0.

Certain optimizations are common. In particular, Depth-First Search prunes paths that contain cycles—namely, pairs of nodes that represent the same state (duplicate nodes), while Breadth-First and Best-First Search keep track of the nodes that have been already selected and expanded in a CLOSED list. Duplicates of these nodes can be pruned except when the new node has a lower accumulated cost $g(n)$ and the search is aimed at returning a minimum-cost path. Similarly, duplicate nodes in the search frontier, also called the OPEN list, are avoided by just keeping in OPEN the node with the least evaluation function.

It can be shown that all these algorithms are *complete*, meaning that if there is a path to a goal node, the algorithms will find a path in finite time.[1] Furthermore, some of these algorithms are *optimal*, meaning that the paths returned upon termination will be optimal. These include Dijkstra's algorithm, Breadth-First Search when action costs are uniform, and A* when the heuristic $h(n)$ is *admissible*, i.e., it doesn't overestimate the true optimal cost $h^*(n)$ from n to the goal for any node n. The complexity in space of these algorithms can be described in terms of the length d of the solutions and the average number of children per node, the so-called *branching factor* b. The space requirement of Breadth-First search is *exponential* and grows with $O(b^d)$, where d is the length of the optimal solution, as a b-ary tree of depth d has b^d leafs all of which can make it into the search frontier in the worst case. The space complexity of A* with admissible heuristics is in turn $O(b^{C^*/c_{\min}})$, where C^* is the optimal cost of the problem and $c_{\min}$ is the minimum action cost. This is because A* may expand in the worst case all nodes n with evaluation function $f(n) \leq C^*$ and such nodes can be at depth $C^*/c_{\min}$.[2] The same bounds apply for the time complexity of the algorithms. On the other hand, the space requirements for Depth-First Search are minor: they grow linearly with d as $O(bd)$, as the search frontier in DFS just needs to keep track of the path to the last selected node along with the children of the ancestors nodes that have not yet been expanded.

The difference between linear and exponential memory requirements can be crucial, as algorithms that require exponential memory may abort after a few seconds or minutes with an "insufficient memory" message. Since DFS is the only linear space algorithm above, extensions of DFS have been developed that use linear memory and yet return optimal solutions. The core of some of these algorithms is a *bounded-cost* variant of DFS where selected nodes n whose evaluation function $f(n) = g(n)$ exceeds a given bound B are immediately pruned. Bounded-Cost DFS remains complete provided that the bound B is not smaller than the cost C^* of an optimal solution, but it is not optimal unless B is equal to C^*. Two iterative variants of Bounded-Cost DFS achieve optimality by performing several successive trials with increasing values for the bound B, until $B = C^*$. Iterative Deepening (ID) is a sequence of Bounded-Cost DFS searches with the bound B_0 for the first iteration set to 0, and the bound B_i for iteration $i > 0$ set to minimum evaluation function $f(n)$ over the nodes pruned in the previous iteration so that at least a new node is expanded in each iteration. In its most standard form, when all action costs are equal to 1, the bound B_i in the iteration i is $B_i = i$. ID

---

[1] For DFS to be complete, paths containing cycles must be pruned to avoid getting trapped into a loop.

[2] This bound assumes that the heuristic is consistent, as otherwise, nodes may have to be reopened an exponential number of times in the worst case [Pearl, 1983].

combines the best elements of Depth-First Search (memory) and Dijkstra's algorithm (optimality). ID achieves this combination by performing multiple searches where the same node may be expanded multiple times. Yet, asymptotically this does not affect the time complexity that is dominated by the worst-case time of the last iteration. Iterative Deepening A* (IDA*) is a variant of ID that uses the evaluation function of A*—namely, $f(n) = g(n) + h(n)$. As long as the heuristic is admissible, the first solution encountered by IDA* will be optimal too, while usually performing fewer iterations than ID and pruning many more nodes [Korf, 1985].

The number of nodes expanded by heuristic search algorithms like A* and IDA* depends on the quality of the heuristic h. A* is no better than Breadth-First Search or Dijkstra's algorithm when $h(n)$ is uniformly 0, and IDA* is no better then than ID. Yet if the heuristic is optimal, i.e., $h(n) = h^*(n)$ where h^* stands for the optimal cost from n to the goal, then both A* and IDA* will find an optimal path to the goal with no search at all, just expanding the nodes in one optimal path only (for this though, A* must break ties in the evaluation function by favoring the nodes with the smaller heuristic, else if the problem has multiple optimal solutions, A* may keep switching from one optimal path to another). In the middle, A* and IDA* will expand fewer nodes using an admissible heuristic h_1 than using an admissible heuristic h_2 when h_1 is higher than h_2.[3] A more common situation is when h_1 is higher than h_2 over some states, and equal to h_2 over the other states. In such cases, h_1 will produce no more expansions than h_2 provided that ties are broken in the same way in the two cases. In the first case, the heuristic h_1 is said to be *more informed* than h_2 or to *dominate* h_2, in the second, that it is at least *as informed as* h_2 [Edelkamp and Schrödl, 2012, Pearl, 1983].

Admissible heuristics are crucial in algorithms like A* and IDA* for ensuring that the solutions returned are optimal, yet they are not crucial for finding solutions fast, when there is no need for optimality. Indeed, these algorithms will often find solutions faster by multiplying an admissible heuristic h by a constant $W > 1$ as in WA*. WA* can be thought as an A* algorithm but with heuristic $W \cdot h$ which is not necessarily admissible even when h is. The reason that WA* will find solutions faster than A* can be seen by considering the nodes selected for expansion given an OPEN list that contains one node n that is deep in the graph but close to the goal, e.g., $g(n) = 10$ and $h(n) = 1$, and another node n' that is shallow and far from the goal; e.g., $g(n') = 2$ and $h(n') = 6$. Among the two nodes, A* will choose n' for expansion as $f(n') = g(n') + h(n') = 2 + 6 = 8$, while $f(n) = g(n) + h(n) = 10 + 1 = 11$. On the other hand if $W = 2$, WA* chooses the node n instead as $f(n') = g(n') + W \cdot h(n') = 2 + 2 \cdot 6 = 14$ and $f(n) = g(n) + W \cdot h(n) = 10 + 2 \cdot 1 = 12$.

The optimality of A* with admissible heuristics can be shown by contradiction. First, it is not hard to verify that until termination, the OPEN list always contains a node n in an optimal path to the goal if the problem is solvable. Then, if A* selects a goal node n' with cost $f(n')$ that is higher than the optimal cost C^*, then $f(n') > C^* \geq g(n) + h(n) = f(n)$ as the heuristic h is admissible. Therefore, A* did not select the node with minimum f-value from OPEN which is a contradiction. WA* is not optimal, even when the heuristic h is admissible, yet the same argument can be used to show that the solution returned by WA* will not exceed the optimal cost C^* by more than a factor of W. This is important in practice, as for example, with $W = 1.2$ the algorithm may turn out to run an order-of-magnitude faster and with much less memory than A*, yet the loss in optimality is then at most 20%.

Anytime WA* [Hansen and Zhou, 2007] is an *anytime optimal* algorithm which basically works exactly like WA* until a solution is found with cost C, not necessarily optimal. Rather than stopping

---

[3]Technically speaking, this result assumes that both heuristics are consistent [Pearl, 1983].

there, however, anytime WA* uses the amount of time available for improving the quality of this solution, continuing the WA* search, while pruning nodes n with accumulated costs $g(n)$ greater than C, and updating the bound C to C' when solutions with cost C' less than C are found. Anytime WA* can thus produce solutions more quickly, and if given enough time, produce better and better solutions until finding an optimal solution. This, however, can only be verified when the search terminates, i.e., when the OPEN list becomes empty. Another interesting anytime optimal algorithm obtained as a variation of WA* is Restarting WA* or RWA* that performs iterated WA* searches but with decreasing weights, while keeping in memory nodes expanded in previous iterations that are re-expanded only when a cheaper path to the node is found [Richter et al., 2010]. This is the search algorithm used in the state-of-the-art heuristic search planner called LAMA [Richter and Westphal, 2010]. Many other heuristic search planners use *Greedy Best-First (GBFS)* which is a best-first search with evaluation function $f(n) = h(n)$, where the accumulated cost term $g(n)$ has been dropped. GBFS can be seen as an WA* algorithm with a very large constant, and a tie breaking rule that favors nodes n with smallest accumulated costs $g(n)$.

2.4   ONLINE SEARCH: THINKING AND ACTING INTERLEAVED

Often there is no time for machines to compute a complete solution offline before performing an action. This is clearly the situation in Chess where programs do not compute a complete solution to the game before choosing a move, but rather focus iteratively on the move to do next. Of course, Chess is a two-player game that cannot be described in terms of classical state models, yet the situation is similar for people in games such as Rubik's Cube, the 15-Puzzle, or Sokoban. People do not have the patience or the computational resources to compute complete solutions offline. Rather, they think a bit, explore the possible options and consequences, and then move. Search algorithms that interleave thinking and acting in this way are called *online* search algorithms. While online search algorithms come often with fewer guarantees than the *offline* algorithms, they are often more practical and have a wider scope. In particular, online search algorithms can deal with classical state models that are not completely accurate, provided that the state of the system is *fully observable*. Indeed, when the model predicts the state s' after performing the action a in the state s, but the state s'' is observed instead, online algorithms can continue the search from s''. Thus, for example, a Blocks World problem where there is a chance for blocks to fall down accidentally from the gripper, can be described in terms of the standard, deterministic Blocks World state model. Then an online search algorithm that keeps track of the state of the system can replan from the new state when it is not the state that was predicted from the model (e.g., where a block fell unexpectedly from the gripper). In the presence of feedback, thus, online search algorithms can be used to provide a form of *closed-loop control* where a bounded search over the model is used to select the action to do in the observed state, which is then applied to the real or simulated system, and so on, until the goal is reached. Online search algorithms are also called *planning and execution* algorithms, as they interleave planning and execution, as opposed to offline search algorithms that just plan.

The simplest heuristic online algorithm is also the simplest heuristic algorithm of all: from the current state s, the action a that is selected is the one that minimizes the estimated cost to the goal defined as:

$$Q(a, s) = c(a, s) + h(s') \tag{2.1}$$

where h is the heuristic function and s' is the state that is predicted to follow the action a in the state s, i.e., $s' = f(a, s)$. The minimization is done over the actions a applicable in s, i.e., $a \in A(s)$.

This algorithm is known as the *greedy algorithm* or *policy*, and also as *hill climbing* search. It's called greedy because it selects the action to be done by trusting the heuristic function h completely, and *hill climbing*, because when action costs are uniform it behaves as if $Q(a, s) = h(s')$, selecting thus actions that minimize the heuristic function toward goal states that should have a zero heuristic.[4]

The main positive property of the greedy algorithm is that it is *optimal* if the heuristic h is optimal, i.e., if $h = h^*$. In addition, the algorithm uses constant memory; it doesn't keep track of a search frontier at all, just the current state and its children. This is however where the good news for the greedy algorithm end. The algorithm in general is neither optimal nor complete; in fact, it can get trapped into a loop, selecting actions that take it from a state s into state s' and then back from s' to s.

A common way to improve the greedy algorithm is by *looking ahead* from the current state s, not just one level as done by the minimization of the $Q(a, s)$ expression in Eq. 2.1, but several levels. A depth-first search from s that prunes nodes that are deeper than a given bound H can be used to perform this lookahead where H is the number of levels, or planning *horizon*. This form of lookahead is time exponential in the horizon, $O(b^H)$, where b is the branching factor of the problem. After this lookahead, the action that is selected is the one on the path to the best leaf, with "best" defined in terms of the evaluation function $f(n) = g(n) + h(n)$.

This form of lookahead ensures that the action selected is the one that is best within the planning horizon H for the given heuristic, yet this horizon must be kept small, else the local lookahead search cannot be completed in real time. A useful alternative lookahead scheme over *small time windows* is the combination of a larger horizon H along with a heuristic search algorithm that operates within this horizon (deeper nodes are still pruned) as an *anytime optimal algorithm*. For example, the lookahead search can be done with the A* algorithm from the current state s. Then, when time is up, whether the search is finished or not, the action selected in s is taken as the one leading to the best leaf, yet with the leaves being both the nodes at depth H that have been generated plus the nodes that have been generated at any other level that have not been yet expanded. Algorithms like Anytime WA* can also be convenient for this type of anytime optimal lookahead search.

While a *depth-first* or *best-first* lookahead can improve the quality of the actions selected in the greedy online search algorithm, neither approach guarantees completeness and optimality, except in the trivial case where a solution and an optimal solution are within an horizon H of the seed state. On the other hand, there is a simple fix to the greedy search that delivers both completeness and optimality. The fix is due to Richard Korf, and the resulting algorithm is known as *Learning Real Time A** or *LRTA** [Korf, 1990].

LRTA* is an extremely simple, powerful, and adaptable algorithm, that as we will see generalizes naturally to MDPs. LRTA* is the online greedy search algorithm with *one change*: once the action a that minimizes the estimated cost-to-go term $Q(a, s)$ from s is applied, the heuristic value $h(s)$ is updated to $Q(a, s)$. The code for LRTA* is shown in Figure 2.2 where the dynamically changing

---

[4]In the planning setting, the algorithm actually does *hill descending*. The name of the algorithm, however, comes from contexts where states that maximize a given function are sought.

LRTA*

% h is the initial value function and V is the hash table that stores the updated
% values. When fetching a value for s in V, if V does not contain an entry for s,
% an entry is created with value h(s)

Let $s := s_0$
while s is not a goal state **do**
    Evaluate each action $a \in A(s)$ as: $Q(a,s) := c(a,s) + V(f(a,s))$
    Select best action **a** that minimizes $Q(\mathbf{a},s)$
    Update value $V(s) := Q(\mathbf{a},s)$
    Set $s := f(a,s)$
end while

Figure 2.2: Single Trial of Learning Real Time A* (LRTA*)

heuristic function, initially set to $h(s)$, is denoted as $V(s)$. For the implementation of LRTA*, the estimates $V(s)$ are stored in a hash table that initially contains the heuristic value $h(s_0)$ of the initial state s_0 only. Then, when the value of a state s that is not in the table is needed, a new entry for s with value $V(s) = h(s)$ is allocated. These entries $V(s)$ are updated as

$$V(s) := \min_{a \in A(s)} Q(a,s) = \min_{a \in A(s)} [c(a,s) + V(s')] \qquad (2.2)$$

where $s' = f(a,s)$, when the action $a = \operatorname{argmin}_{a \in A(s)} Q(a,s)$ is applied in the state s. This simple greedy algorithm combined with these updates delivers the *two key properties* provided that the heuristic $h(s)$ is *admissible* and that there are *no dead-ends* (states from which the goal cannot be reached). First, LRTA* will not be trapped into a loop and will eventually reach the goal. Second, if upon reaching the goal, the search is restarted from the same initial state while keeping the current heuristic function V, and this process is repeated iteratively, eventually LRTA* converges to an optimal path to the goal. This convergence will be achieved in a finite number of iterations, and the convergence is achieved when the updates $V(s) := \min_{a \in A(s)} Q(a,s)$ do not change the value $V(s)$ of any of the states encountered in the way to the goal, which are then optimal.

These are two remarkable properties that follow from a simple change in the greedy algorithm that adjusts the value of the heuristic according to Eq. 2.2 over the states that are visited in the search. Of course, LRTA*, unlike the greedy algorithm, does not run in constant space, as the updates to the heuristic function take space in the hash table that in the worst case can become as large as the number of states in the problem. The value of the initial heuristic is critical in the performance of LRTA*, both in terms of time and space, as better heuristic values mean a more focused search, and a more focused search means more updates on the states that matter. When LRTA* is to be run once and not until convergence, a lookahead can improve the quality of the actions selected and boost the heuristic values of the visited states (which remain admissible if they are initially admissible). The latter can be achieved if the states that are expanded in the lookahead search are also updated using Eq. 2.2, and the new values are propagated up to their parents. In this way, a move from s will leave a heuristic

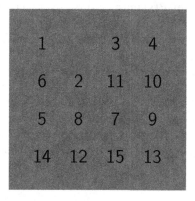

Figure 2.3: The sliding 15-puzzle where the goal is to get to a configuration where the tiles are ordered by number with the empty square last. The actions allowed are those that slide a tile into the empty square. While the problem is not simple, the heuristic that sums the horizon and vertical distances of each tile to its target position is simple to compute and provides informative estimates to the goal. In planning, heuristic functions are obtained automatically from the problem representation.

value $V(s)$ for s that would be more informed than the value of s computed from its children using Eq. 2.2. LSS-LRTA* is a version of LRTA* with a lookahead of this type [Koenig and Sun, 2009].

2.5 WHERE DO HEURISTICS COME FROM?

Heuristic search algorithms express a form of goal-directed search where heuristic functions are used to guide the search toward the goal. A key question is how such heuristics can be obtained for a given problem. A useful heuristic is one that provides good estimates of the cost to the goal and can be computed reasonably fast. Heuristics have been traditionally devised according to the problem at hand [Edelkamp and Schrödl, 2012, Pearl, 1983]: the Euclidean distance is a good heuristic for route finding, the sum of the Manhattan distances of each tile to its destination is a good heuristic for the sliding puzzles, the assignment problem heuristic is good for Sokoban [Junghanns and Schaeffer, 2001], both the assignment problem and spanning tree heuristics have been used for the Travelling Salesman Problem [Lawler *et al.*, 1985], and so on. The general idea that emerges from the various problems is that heuristics $h(s)$ can be seen as encoding the cost of reaching the goal from the state s in a problem that is simpler than the original one [Minsky, 1961, Pearl, 1983, Simon, 1955]. For example, the sum-of-Manhattan distances in the sliding puzzles (Figure 2.3) corresponds to the optimal cost of a simplification of the puzzle where tiles can be moved to adjacent positions, whether these positions are empty or not. Similarly, the Euclidean heuristic for route finding is the cost of a simplification of the problem where straight routes are added between any pair of cities in the map. The simplified problems are normally referred to as *relaxations* of the original problem. If P is the original problem, P' is its relaxation, and $P(s)$ and $P'(s)$ refer to the problem and relaxation when the initial state is set to the state s, the general idea is to set the heuristic value $h_P(s)$ associated with the problem $P(s)$ to the optimal cost $h_{P'}^*(s)$ of the relaxed problem $P'(s)$. It is easy to show that if the solutions to the original problem $P(s)$ are also solutions of the relaxed problem $P'(s)$, something which is natural for most relaxations, then the heuristic $h_P(s)$ that results from the relaxation is actually admissible. This is because any optimal solution for $P(s)$ must be also a solution to the relaxation $P'(s)$, whose optimal value cannot exceed then the optimal value of $P(s)$. On the other hand, if the heuristic $h_P(s)$ for P is obtained from a solution to the relaxation $P'(s)$ that is not necessarily optimal, the resulting heuristic $h_P(s)$ would not be necessarily admissible.

A key development in modern planning research was the realization that useful heuristics could be derived automatically from the representation of the problem in a domain-independent planning language [Bonet et al., 1997, McDermott, 1996]. It does not matter what the problem $P(s)$ is about, an automated relaxation $P'(s)$ yielding informative heuristics can be obtained directly and effectively from the representation of $P(s)$. The result is a *domain-general* heuristic $h(s)$, i.e., a heuristic that makes the search goal-driven, no matter what the problem is about, as long as it is a problem where deterministic actions expressed in compact form have to be used to drive the system from a known initial state into a goal state.

2.6 LANGUAGES FOR CLASSICAL PLANNING

The languages for expressing classical planning models in compact form come in two main varieties. In one, the state variables are all boolean, i.e., they can take just one of two values, true or false. In the other, they are multivalued and can take values from a finite domain. In either case, the states in the resulting model are the valuations over the variables, where a valuation assigns to each variable a value from its domain.

The simplest and possibly oldest classical planning language in use is STRIPS, a language based on boolean variables, which was originally developed in a different form for controlling the Robot Shakey at SRI during the late 60s [Fikes and Nilsson, 1971]. A planning problem in the current version of STRIPS is a tuple $P = \langle F, I, O, G \rangle$ where

- F represents the set of *atoms* or *propositions* of interest,

- O represents the set of *actions*,

- $I \subseteq F$ represents the *initial situation*, and

- $G \subseteq F$ represents the *goal*.

In STRIPS, the actions $o \in O$ are represented by three sets of atoms over F called the Add, Delete, and Precondition lists, denoted as $Add(o)$, $Del(o)$, $Pre(o)$. The first describes the atoms that the action o makes true, the second, the atoms that o makes false, and the third, the atoms that must be true in order for the action to be applicable. A STRIPS problem $P = \langle F, I, O, G \rangle$ encodes implicitly, in compact form, the classical state model $\mathcal{S}(P) = \langle S, s_0, S_G, A, f, c \rangle$ where

- the states $s \in S$ are the possible *collections of atoms* over F, each defining a truth valuation where an atom $p \in F$ is true in s iff $p \in s$,

- the initial state s_0 is I,

- the set S_G of goal states comprises the states s for which $G \subseteq s$,

- the actions a in $A(s)$ are the ones in O with $Prec(a) \subseteq s$,

- the state transition function is $f(a, s) = (s \setminus Del(a)) \cup Add(a)$, so that the state s' that results from action a in s is s but with the atoms in $Del(a)$ deleted, and the atoms in $Add(a)$ added, and

- the action costs $c(a, s)$ are equal to 1 by default.

Given that the STRIPS problem P represents the state model $\mathcal{S}(P)$, the *plans* for P are defined as the plans for $\mathcal{S}(P)$, namely, the action sequences that map the initial state s_0 that corresponds to I into a goal state where the goals G are true. Since the states in $\mathcal{S}(P)$ are represented as collections of atoms from F, the number of states in $\mathcal{S}(P)$ is $2^{|F|}$ where $|F|$ is the number of atoms in P, usually called *fluents*.

The state representation that follows from a planning language such as STRIPS is domain-independent. Thus, while a specialized solver for a Blocks World problem may represent the state of the problem by a set of lists, each one representing a tower of blocks, in the state representation that follows from STRIPS there are just atoms, and the same is true of any other domain. As an illustration, a domain that involves three locations l_1, l_2, and l_3, and three tasks t_1, t_2, and t_3, where t_i can be performed only at location l_i, can be modeled with a set F of fluents $at(l_i)$ and $done(t_i)$, and a set O of actions $go(l_i, l_j)$ and $do(t_i)$, $i, j = 1, \ldots, 3$, with precondition, add, and delete lists

$$Pre(a) = \{at(l_i)\}, \ Add(a) = \{at(l_j)\}, \ Del(a) = \{at(l_i)\}$$

for $a = go(l_i, l_j)$, and

$$Pre(a) = \{at(l_i)\}, \ Add(a) = \{done(t_i)\}, \ Del(a) = \{\}$$

for $a = do(t_i)$. The problem of doing tasks t_1 and t_2 starting at location l_3 can then be modeled by the tuple $P = \langle F, I, O, G \rangle$ where

$$I = \{at(l_3)\} \ \text{and} \ G = \{done(t_1), done(t_2)\}.$$

A solution to P is an applicable action sequence that maps the state $s_0 = I$ into a state where the goals in G are all true. In this case one such plan is the action sequence

$$\pi = \langle go(l_3, l_1), do(t_1), go(l_1, l_2), do(t_2) \rangle.$$

The number of states in the problem is 2^6 as there are six boolean variables. Still, it can be shown that many of these states are not reachable from the initial state. Indeed, the atoms $at(l_i)$ for $i = 1, 2, 3$ are mutually exclusive and exhaustive, meaning that every state reachable from s_0 by applying the available actions makes one and only one of these atoms true. These boolean variables encode indeed the possible values of the multivalued variable that represents the agent's location.

Planning languages featuring non-boolean variables and richer syntactic constructs than STRIPS are also common in planning [Bäckström and Nebel, 1995, Gerevini et al., 2009]. In particular, if X is a multivalued variable with domain D_X, then the initial situation can be characterized by a set of literals of the form '$X = x$" for each variable X where $x \in D_X$, the actions can be described in terms of pre and postconditions expressed through these literals, and the same for goals. In principle, a planning problem expressed through multivalued variables can be compiled automatically into a problem with boolean variables only, by simple transformations such as replacing each literal $X = x$

by the proposition $p(X = x)$ throughout, and by including the propositions $p(X = x')$ in the delete list of every action that adds $p(X = x)$. Planning problems expressed over boolean variables such as STRIPS can be similarly expressed in multivalued form by mapping atoms p into literals $X_p = true$ throughout, except in delete lists when they are mapped into postconditions $X_p = false$. Often it is possible to derive a more compact multivalued encoding of a planning problem, e.g., like when a set of atoms $at(l_1)$, ..., $at(l_n)$ is used to represent the possible locations of an object. Such a location can be encoded through a multivalued variable L with domain $D_L = \{l_1, \ldots, l_n\}$. Programs that automatically infer *invariants* from a STRIPS encoding, such as sets of exhaustive and mutually exclusive atoms, are used to transform one representation into another automatically [Helmert, 2009]. While some of the representation languages are more natural for users, they are not necessarily more efficient for planning as the extra features can often be compiled away at no cost. For example, STRIPS does not accommodate *negation* or *negated atoms*, yet when it is convenient to introduce a negated atom $\neg p$ in the initial situation, preconditions, or goals of a problem, it is possible to introduce a new atom $\bar{p}$ for capturing $\neg p$. The atom $\bar{p}$ has to be part of I when $p \notin I$, has to be included in the Add list of an action when p is included in the Delete list, and vice versa, has to be included in the Delete list when p is in Add list. Then the atom $\bar{p}$ can be used in preconditions and goals, as indeed, $\bar{p}$ represents $\neg p$ over all the states that are reachable from I using the available actions, where the logical formula $\bar{p} \equiv \neg p$ can be shown to hold. In other words, the formula $\bar{p} \equiv \neg p$ is an *invariant* in the problem.

One important syntactic construct that extends STRIPS and is not convenient to compile away in general, because of a potential exponential blow up in the size of the problem, is *conditional effects* [Gazen and Knoblock, 1997, Nebel, 2000]. While Add and Delete lists represent sets of atoms that become true and false *unconditionally* after an action is done, a conditional effect $C \to C'$ associated with an action, where C and C' are sets of literals (atoms or negated atoms), says that C' will be true right after the action if C was true right before the action. In other words, unlike an action precondition, C does not have to be true for the action to be applicable, yet if *it is true*, then C' will become true as a result of the action.

Figure 2.4 shows a description of the Blocks World domain in PDDL. PDDL is the Planning Domain Definition Language, a language and syntax that has been used in the planning competitions [McDermott et al., 1998]. PDDL accommodates the STRIPS language along with a number of additional syntactic constructs in a notation that originates in the Lisp programming language. Problems in PDDL are expressed in two parts: one about the general domain; the other about a particular domain instance. In the domain part, the actions are described by means of schemas over generic atoms defined using predicates names like $clear$, variables like $?x$, and possibly constants. In the instance part, the object names that will replace the variables are declared, along with the atoms describing the initial state, and the formula describing the goal states. The "requirement" flag in the domain definition describes the PDDL fragment used by the encoding which can include STRIPS, ADL extensions featuring negation, conditional effects, function symbols, and various forms of quantification [Pednault, 1989], a hierarchy of types for controlling how variables can be substituted by object names, the equality predicate, and so on. There are currently tens of classical planners that can be downloaded free from the Internet and hundreds of planning problems expressed in PDDL for use with such planners.

```
(define (domain BLOCKS)
    (:requirements :strips)
    (:predicates (clear ?x) (on-table ?x) (arm-empty) (holding ?x) (on ?x ?y))
    (:action stack
        :parameters (?x ?y)
        :precondition (and (holding ?x) (clear ?y))
        :effect (and (not (holding ?x)) (not (clear ?y)) (clear ?x) (on ?x ?y) (handempty)))
    (:action unstack
        :parameters (?x ?y)
        :precondition (and (clear ?x) (on ?x ?y) (handempty))
        :effect (and (not (clear ?x)) (not (on ?x ?y)) (not (handempty)) (holding ?x)
                (clear ?y)))
    (:action put_down
        :parameters (?x)
        :precondition (holding ?x)
        :effect (and (not (holding ?x)) (clear ?x) (ontable ?x) (handempty)))
    (:action pick_up
        :parameters (?x)
        :precondition (and (clear ?x) (ontable ?x) (handempty))
        :effect (and (not (clear ?x)) (not (ontable ?x)) (not (handempty)) (holding ?x)))
)

(define (problem BLOCKS_6)
    (:domain BLOCKS)
    (:objects A B C D E F)
    (:init (clear B) (clear C) (clear E) (ontable C) (ontable D) (on A D) (on B A)
        (ontable F) (on E F) (handempty))
    (:goal (and (on E F) (on F C) (on C B) (on B A) (on A D)))
)
```

Figure 2.4: A Blocks World instance described in PDDL.

2.7 DOMAIN-INDEPENDENT HEURISTICS AND RELAXATIONS

A STRIPS planning problem $P = \langle F, I, O, G \rangle$ defines a state model and a directed graph so that the plans for P corresponds to paths connecting a source node to a target node in the graph. The size of the graph, however, is exponential in the number of atoms in P, and this prevents the use of blind search methods in general. As mentioned, a key development in modern planning research was the realization that this search could be guided by *heuristics* extracted automatically from the problem [Bonet and Geffner, 2001, McDermott, 1999]. The heuristics are derived from relaxations, and the most common and useful domain-independent relaxation in planning is the *delete-relaxation*, that maps a problem $P = \langle F, I, O, G \rangle$ into a problem $P^+ = \langle F, I, O^+, G \rangle$ that is exactly like P but with the actions in O^+ set to the actions in O with empty delete lists. That is, the delete-relaxation is a domain-independent relaxation that takes a planning problem P and produces another problem P^+ where atoms are added exactly as in P but they are never deleted. The relaxation implies, for example, that objects and agents can be in "multiple places" at the same time as when an object or an agent moves into a new place, the atom representing the old location is not deleted. Relaxations, however, are not aimed at providing accurate models of the world; quite the opposite, simplified and even meaningless models of the world that while not accurate yield useful heuristic guidance.

The domain-independent delete-relaxation *heuristic* is obtained as an approximation of the optimal cost of the delete-relaxation P^+ obtained from the cost of a plan that solves P^+ not necessarily optimally. The reason that an approximation is needed is because finding an optimal plan for a delete-free planning problem like $P^+(s)$ is still a computationally intractable task (also NP-hard). On the other hand, finding just one plan for the relaxation whether optimal or not, can be done quickly and efficiently. The property that allows for this is *decomposability:* a problem without deletes is decomposable in the sense that a plan π for a joint goal G_1 and G_2 can always be obtained from a plan π_1 for the goal G_1 and a plan π_2 for the goal G_2. The concatenation of the two plans $\pi = \pi_1, \pi_2$ in either order is indeed one such plan. This property allows for a simple method for computing plans for the relaxation from which the heuristics are derived.

The main idea behind the procedure for computing the heuristic $h(s)$ for an arbitrary planning problem P can be explained in a few lines. For this, let $P(s)$ refer to the problem that is like P but with the initial situation set to the state s, and let $P^+(s)$ stand for the delete-relaxation of $P(s)$, i.e., the problem that is like $P(s)$ but where the delete-lists are empty. The heuristic $h(s)$ is computed from a plan for the relaxation $P^+(s)$ that is obtained using the decomposition property and a simple iteration. Basically, the plans for achieving the atoms p that are already true in the state s, i.e., $p \in s$, are the empty plans. Then if $\pi_1, \pi_2, \ldots, \pi_m$ are the plans for achieving each of the preconditions $p_1, p_2, \ldots, p_m$ of an action a that has the atom q in the add list, $\pi = \pi_1, \pi_2, \ldots, \pi_m$ followed by the action a is a plan for achieving q from s. It can be shown that this iteration yields a plan in the relaxation $P^+(s)$ for each atom p that has a plan in the original problem $P(s)$ in a number of steps that is bounded by the number of atoms in the problem. A plan for the actual goal G of $P(s)$ in the relaxation $P^+(s)$ can then be obtained in a similar manner by concatenating the plans for each of the atoms q in G in any order. Such a plan for the relaxation $P^+(s)$, denoted as $\pi^+(s)$, is called a *relaxed plan*. The heuristic $h(s)$ can then be set to the cost of such a plan. A better estimate can be obtained if duplicate actions in the resulting relaxed plan are removed, since no STRIPS action needs to be done twice for solving a delete-free problem, as the effects of the first action occurrence stay true until the end of the plan.[5] Below we will formalize the domain-independent planning heuristic sketched above, and explore some variations.

As an illustration, Figure 2.5 displays a fragment of the directed graph corresponding to a blocks world problem P with the automatically derived heuristic values next to some of the nodes. The heuristic values shown are computed very fast, in low polynomial time, using an algorithm similar to the one described above, with $h(s)$ representing an approximation of the number of actions needed (cost) to solve the relaxed problem $P^+(s)$. Actually, the instance shown can be solved without any search at all by just selecting in each node, starting from the source node, the action that leads to the node with a lower heuristic value (closer to the goal). The resulting plan is shown as a red path in the figure.

---

[5]This is not true, however, in planning languages that extend STRIPS with negation and conditional effects, where the same action may have to be applied multiple times for solving the relaxation.

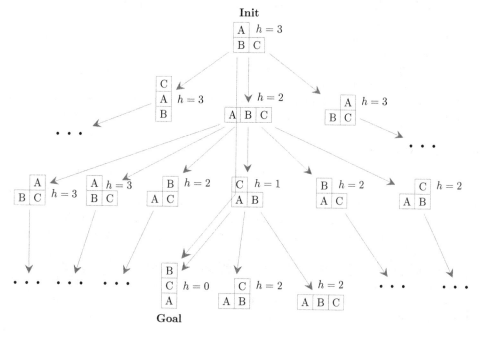

Figure 2.5: A fragment of the graph corresponding to a blocks world planning problem with the automatically derived heuristic values shown next to some of the nodes. The heuristic values are computed in low polynomial time and provide the search with a sense of direction. The instance can actually be solved without any search by just performing in each state the action that leads to the node with a lower heuristic value (closer to the goal). The resulting plan is shown in red; helpful actions shown in blue.

In order to get a more vivid idea of where the heuristic values shown in the figure come from, consider the heuristic $h(s)$ for the initial state where block A is on B, and both B and C are on the table. In order to get the goal "B on C" in the relaxation from the state s, two actions are needed: one to get A out of the way to achieve the preconditions for moving B, the second to move B on top of C. On the other hand, in order to achieve the second goal "C on A" in the relaxation from s, just the action of moving C to A is needed. The result is a heuristic value $h(s) = 3$ as shown, which actually coincides in this case with the cost of the best plan to achieve the joint goal from s in the non-relaxed problem $P(s)$. Nonetheless, this is just a coincidence, and indeed, the best plans in the relaxation $P^+(s)$ can be quite different than the best plans in the original problem $P(s)$. The best plan for $P(s)$ is indeed unique: moving A to the table, then C on A, and finally B on C. On the other hand, a possible optimal plan in the relaxation $P^+(s)$ is to move first C on A, then A on the table, and finally B on C. Of course, this plan does not make any sense in the real problem where A can't be moved when covered by C, yet the relaxation is not aimed at capturing the real problem or the real physics; it is aimed at producing informative but quick estimates of the cost to the goal. The reader can verify that for the leftmost child s' of the initial state s, the costs of the problem $P(s')$ and the relaxation $P^+(s')$ no longer coincide. The former is 4, while the latter is 3, the difference arising from the goal "C on A" that in the original problem must be undone and then redone. In the relaxation this is never needed as no atom is ever deleted.

The heuristics for classical planning that have been developed so far, all assume that actions costs $c(a, s)$ are 1 or depend at most on the action a but not on the state s, i.e., $c(a, s) = c(a)$. In principle, there is no problem in expressing arbitrary action costs $c(a, s)$ in compact form, like for instance saying that $c(a, s)$ is 1 except that when s makes both p and q true where it is 100. Yet, while such cost structures are important and are often needed, they have not been addressed systematically in the literature so far, and thus there are not yet good heuristics for handling them in planning.

ADDITIVE AND MAX HEURISTICS

One of the first heuristics developed for domain-independent planning operate on the delete-relaxation but doesn't involve the computation of relaxed plans. We review this heuristic below along with some variations that actually do. In order to simplify the definition, we introduce a new dummy End action with *zero cost*, whose preconditions $G_1, \ldots, G_n$ are the goals of the problem, and whose effect is a dummy atomic goal G. The heuristics $h(s)$ simply estimates the cost of achieving this "dummy" goal G from the state s.

Since the heuristic h^+ that represents the optimal cost function of the delete-relaxation is intractable, the *additive heuristic h_{add}* introduces a polynomial approximation in which subgoals are assumed to be *independent* in the sense that they are achieved with no "side effects" [Bonet et al., 1997]. This assumption is normally false, but results in a simple heuristic function

$$h_{add}(s) \stackrel{\text{def}}{=} h_{add}(Pre(End); s) \tag{2.3}$$

that can be computed quite efficiently in every state s visited in the search, where $h_{add}(Pre(a); s)$ is an estimate of the cost of achieving the preconditions of action a from s, defined from the expressions:

$$h_{add}(p; s) \stackrel{\text{def}}{=} \begin{cases} 0 & \text{if } p \in s \\ \min_{a \in O(p)} [cost(a) + h_{add}(Pre(a); s)] & \text{otherwise} \end{cases} \tag{2.4}$$

and

$$h_{add}(Pre(a); s) \stackrel{\text{def}}{=} \sum_{q \in Pre(a)} h_{add}(q; s). \tag{2.5}$$

In these expressions, $h_{add}(p; s)$ stands for the estimated cost of achieving the atom p from s, $O(p)$ stands for the actions in the problem that add p, and $h_{add}(Pre(a); s)$ stands for the estimated cost of achieving the preconditions of the actions a from s. Versions of the additive heuristic appear in several planners [Bonet and Geffner, 2001, Do and Kambhampati, 2001, Smith, 2004], where the cost of the joint condition in action preconditions (and goals) is set to the sum of the costs of each condition in isolation. The additive heuristic h_{add} is neither a lower bound nor an upper bound on the optimal cost function h^* over the original problem. The reason is that the cost of achieving two atoms jointly from a state s can be lower or higher than the sum of the costs of achieving each one of them individually. In particular, if a is a unit-cost action with preconditions that are true in s and atoms p and q in the Add list that are not true in s, then the heuristic $h_{add}(s)$ for the goal $G = \{p, q\}$ will be 2, while clearly the optimal cost $h^*(s)$ of achieving G in the problem from s is 1. Likewise, if there is instead just one action that adds p and deletes q, and one action that adds q and deletes p, then the heuristic $h_{add}(s)$ would still be 2, while the optimal cost $h^*(s)$ of achieving G in the problem would be infinity. G is indeed unachievable in such a problem where the formula $\neg(p \wedge q)$ is an invariant.

If the estimated cost of the joint condition in Eq. 2.5 is changed from the *sum* to the *maximum*,

$$h_{max}(Pre(a); s) \stackrel{\text{def}}{=} \max_{q \in Pre(a)} h_{max}(q; s), \tag{2.6}$$

a different heuristic is obtained by setting $h_{max}(s)$ to $h_{max}(Pre(End); s)$, and replacing h_{add} by h_{max} in Eq. 2.4. Since the cost of achieving several atoms from a state s can never be lower than the cost of achieving one of them, the max-heuristic h_{max} unlike the additive heuristic h_{add} is *admissible*, and hence, potentially useful for computing optimal plans in combination with algorithms such as A* or IDA*. Still, the max-heuristic is not informative enough in general, as it ignores all but one of the atoms of each action precondition. In this sense, the heuristic h_{add} is not admissible but is better at discriminating good from bad actions, as no precondition is left out from the computation.

The equations for h_{add} and h_{max} basically define a path-finding problem over *atom space* as opposed to the planning problem that is a path-finding problem over the exponentially larger *state space*. Indeed, any shortest-path algorithm can be used for computing these heuristics, including Dijkstra's algorithm, Bellman and Ford's, or Value Iteration [Bertsekas, 1995, Cormen et al., 2009]. A single change is needed though: while the nodes of the graph are the problem atoms, the edges, that correspond to problem actions are actually *hyperedges* rather than normal *edges*, as they link the set of atoms appearing in an action precondition with each of the atoms appearing in the Add list. So, while an edge (n, n') in a normal directed graph induces a cost $c(n') \leq c(n) + w(n, n')$ on the target node n', a (directed) hyperedge $(\{n_1, \ldots, n_k\}, n')$ associated with an action a induces a cost $c(n') \leq \sum_{i=1,k} c(n_i) + c(a)$ instead, in the additive heuristic. For any of the algorithms, the costs $c(n)$ can be initialized to 0 for the nodes n corresponding to atoms p that are true in s, and to ∞ for all other atoms. All the algorithms use the inequalities $c(n') \leq \sum_{i=1,k} c(n_i) + c(a)$ as updates of the form $c(n') := \min(c(n'), \sum_{i=1,k} c(n_i) + c(a))$, and differ in the order in which these updates are performed and the conditions under which they are terminated. Dijkstra's algorithm for example updates nodes n', once, in order, according to the value of the right-hand side expression, lowest first. Bellman and Ford's algorithm, and Value Iteration, do not pay the overhead required for following this order, but end up updating nodes many times. In all cases, the computation is polynomial and finishes in Dijkstra's algorithm when there are no more nodes to update, while in Bellman and Ford's algorithm and in Value Iteration, when the updates produce no changes. For the max heuristic, the update expression needs to be changed to $c(n') := \min(c(n'), \max_{i=1,k} c(n_i) + c(a))$. An analysis of these various methods for computing the additive heuristic is given by Liu et al. [2002].

RELAXED PLAN HEURISTIC

The heuristic search planners UNPOP and HSP use the additive heuristic in the context of standard heuristic search algorithms [Bonet and Geffner, 2001, McDermott, 1999]. The planner FF that built on these planners introduced two important changes [Hoffmann and Nebel, 2001]: one in the heuristic, and one in the search procedure. We focus here on the heuristic, and consider the search procedure in the next chapter. FF's heuristic $h_{FF}(s)$ is set to the cost of a plan $\pi_{FF}(s)$ for the relaxation $P^+(s)$ that is not necessarily optimal, and which is obtained by running a Graphplan-like procedure [Blum and Furst, 1995] that exploits the decomposability of the delete-free problem in line with the algorithm sketched above. Basically, a graph made of successive layers $P_0, A_0, P_1, A_1, \ldots$ where P_i is a set of atoms, and A_i is a set of actions, is constructed for a problem $P(s) = \langle F, I = s, O, G \rangle$ as:

$$P_0 = \{p \in s\}$$
$$A_i = \{a \in O \mid Pre(a) \subseteq P_i\}$$
$$P_{i+1} = P_i \cup \{p \in Add(a) \mid a \in A_i\}$$

until a fixed point is reached; i.e., a layer P_n for which $P_{n+1} = P_n$. Here P_0 contains all the atoms in s, A_i contains all the actions whose preconditions are true in P_i, and P_{i+1} contains the positive effects of these actions along with the atoms appearing in previous layers $P_k, k \leq i + 1$. The resulting layered graph cannot contain more than $|F|$ layers, which happens only when P_0 is empty, and P_{i+1} just contains one more atom than P_i. Moreover, the construction can be stopped when the goal G first appears in a layer P_{m+1}, i.e., $G \subseteq P_{m+1}$, as the plan $\pi_{FF}(s)$ for the relaxation $P^+(s)$ is extracted then backward from that layer. For this, it's convenient to conceive $\pi_{FF}(s)$ as a "parallel" plan made up of actions B_0 done in "parallel" at time 0, actions B_1 done in "parallel" at time 1, and so on until B_m. By actions done in parallel, we mean that the actions can be done in any order, as they will necessarily have the same effect in the relaxation. In addition, during the construction of the graph, each atom p that makes it for the first time in a layer P_{i+1} is tagged with one of the actions in A_i that adds p which is called the *best supporter* for p and is denoted by a_p. Clearly, there must be one such action, else the atom p would not make it into the layer P_{i+1}. The sets of actions B_i in the relaxed plan $\pi_{FF}(s)$ is then obtained recursively backward from the set G_{i+1} of atoms, initially set to G for $i = m$. Then for $i = m, m - 1, \ldots, 0$, and starting with $B_i = \emptyset$, we add to B_i, the best supporter a_p of each of the atoms p in G_{i+1} that made into the layer P_{i+1} for the first time, and recursively set G_i to $(G_{i+1} \setminus Add(B_i)) \cup Pre(B_i)$, where $Add(B_i)$ and $Pre(B_i)$ are respectively the union of the Add and Precondition lists of the actions in B_i.

It is easy to see that the resulting plan $\pi_{FF}(s)$ made up of this sequence of action sets B_i, where actions in a set can be done in any order, is a plan for the relaxation $P^+(s)$. This is because $\pi_{FF}(s)$ contains actions that add each of the goals in G, and in addition, every action in $\pi_{FF}(s)$ has preconditions that are true in the state s or are added by previous actions in the sequence. The heuristic $h_{FF}(s)$ is defined as the *size* $|\pi_{FF}(s)|$ of the plan, namely, the number of actions that it contains, thus assuming implicitly that action costs are all 1. Of course, $h_{FF}(s)$ could be defined instead as the sum of the action costs $c(a)$ for a in $\pi_{FF}(s)$, yet this does not address the fact that the relaxed plan was constructed assuming that costs were uniform.

An advantage of FF's heuristic over the additive heuristic is that it is less likely to overcount actions. For example, if there is an action a whose preconditions hold in s with effects p and q that do not, then the additive heuristic for the goal $G = \{p, q\}$ is 2, while FF's heuristic will be 1. Still, FF's heuristic can also overcount if there is a second action that adds q and whose preconditions hold in s. In such a case, FF can produce a relaxed plan where the atom p is supported by the first action, and the atom q is supported by the second action. An additional limitation of the FF heuristic is that it assumes that all action costs are uniform. There is however a simple way to combine the benefits of the additive and FF's heuristic [Keyder and Geffner, 2008a]. For this, all that is required is to change the definition of the *best support* action a_p for each atom p in the computation of the layered graph to

$$a_p \overset{\text{def}}{=} \operatorname{argmin}_{a \in O(p)} [c(a) + h_{add}(Pre(a); s)] \tag{2.7}$$

when p is not true in the state s. These best supports obtained from the additive heuristic can then be used to build a relaxed plan that no longer ignores action costs. The backward procedure for extracting a plan from these best supports proceeds like above: we collect in the relaxed plan the best supports for the atoms in the goal, and recursively, the best supports of their preconditions, in all cases skipping the atoms that are true in the state s. Actually, the same construction can be used also with the max heuristic. Interestingly, the relaxed plan that would be obtained in this way from the max heuristic is *equivalent* to the one obtained from FF's procedure, provided that ties in the selection of best supports are broken in the same way. This is because there is a tight correspondence between the max heuristic and the layered graph constructed by FF. Indeed, it can be easily shown that the heuristic $h_{max}(s)$ is equal to the index i of the first propositional layer P_i that contains the goal G of the problem [Haslum and Geffner, 2000]. This also implies that the construction of the layered graph provides an alternative way for computing the h_{max} heuristic that is still polynomial, although taking more space than the shortest-path formulation over atom space described above.

2.8 HEURISTIC SEARCH PLANNING

The first generation of heuristic search planners in the late 90s looked for plans by plugging one of these heuristics into a standard heuristic search algorithm. HSP, in particular, used the additive heuristic inside a WA* search. The planner FF, on the other hand, used the relaxed planning heuristic $h_{FF}(s)$ and the relaxed plan $\pi_{FF}(s)$ itself in a search architecture that works in two phases. In the first phase, an incomplete but fast search called Enforced Hill Climbing (EHC) is performed. If a plan is found, it is reported, else a second phase is triggered consisting of a complete Greedy Best First Search guided by the h_{FF} heuristic. This second phase is not too different from HSP as the search algorithm and the heuristics are similar to the ones used in HSP. The main novelty of FF is in the first phase. First, the EHC search ignores all the actions a that are not "helpful" in a given state, where an action is deemed as helpful in a state s when it is applicable in s and adds a goal or the precondition of an action in the relaxed plan $\pi_{FF}(s)$ that is not true in s. Second, with the actions pruned in this way, thus reducing significantly the branching factor of the problem, a Breadth-First Search is triggered from s until a state s' is found with a heuristic value $h_{FF}(s')$ that is strictly smaller than $h_{FF}(s)$. The actions leading from s to s' are then applied, and the process is repeated from s' until a state s'' is found with $h_{FF}(s) = 0$, which is necessarily a goal state. The EHC search is thus a Greedy search with a lookahead restricted to the actions that are helpful over an horizon that is defined implicitly by the presence of a state with a smaller heuristic value. The EHC search is extremely fast in problems with high branching factors where standard best-first search is unfeasible or too slow. More recent planners like Fast Downward and LAMA [Helmert, 2006, Richter and Westphal, 2010] have managed to improve on FF by incorporating "helpful actions" inside a complete search scheme. In FF, on the other hand, if the incomplete EHC search fails, the useful distinction between "helpful" and "non-helpful" actions is lost. We will come back to these issues in the next chapter.

These planners can scale up to very large problems, but they make no attempt at computing provable optimal solutions. Their performance is assessed empirically by comparing the number of problems that they solve, the time that they take, and the quality of the solutions that they find. Additional information, like the number of nodes that are expanded in the search, provides an indication of how informative are the heuristics and how selective is the search.

2.9   DECOMPOSITION AND GOAL SERIALIZATION

The domain-independent heuristics developed for planning often yield quite impressive results, but can also fail miserably sometimes. It is not surprising when the heuristics fail on domains that are inherently hard [Hoffmann et al., 2007], yet the heuristics can fail on trivial domains too. VISIT-ALL is a domain from the 2011 Int. Planning Competition [Coles et al., 2012] where an agent in the middle of a square $n \times n$ grid must visit all the cells in the grid. This is an extremely simple problem to solve when there are no optimality requirements, yet planners such as HSP and FF do not scale up to values of n greater than 20. Indeed, a Greedy Best-First search guided by the additive heuristic does a memory out for $n = 15$, and produces plans that are much longer than optimal, at cost 883 for $n = 10$. The planner LAMA, in the default configuration which uses landmarks—to be explained below—does much better in this domain, solving grids with n up to 30, with relatively good plans. Indeed, for $n = 10$, the solution found by LAMA has cost 138, much lower than 883. The problem with the additive heuristic in this domain is that when the agent gets closer to one goal, visiting one particular cell, it gets away from other goals. Thus, huge plateaus must be traversed in the search before the additive heuristic can be decreased. This causes problems both in best-first search algorithms, and in the otherwise quite effective Enforced Hill Climbing (EHC) search. Actually, VISIT-ALL is a domain that is simple because it's easy to decompose and solve, just requiring to visit the closest unvisited cell, one at a time, until all cells have been visited. Indeed, there is a very simple domain-independent heuristic that does not look at the actions in the domain at all, and yet does better than heuristics like $h_{add}(s)$ and $h_{FF}(s)$ that do: it's the *number of unachieved goals* heuristic $h_{ug}(s)$ that simply counts the number of top goals in the problem, i.e., atoms in G, that are not true in the state s. A Greedy Best-First search planner with this heuristic does actually better than LAMA and the other planners, solving easily problems with n greater than 50, producing good plans too. For $n = 10$, the resulting plan has cost 104, smaller than LAMA's 138, and much smaller than the 883 steps that result from a GBFS search guided by additive heuristic.

There are several lessons to draw from this simple example. First, that heuristics like h_{add} and h_{FF} are not necessarily better than simpler heuristics like h_{ug}, which simply counts the number of unachieved goals, without trying to estimate the cost of achieving them. Second, and related to the first point, heuristics like h_{add} and h_{FF} fail often to decompose a problem into subproblems, even when this appears trivial and effective. Of course, the number of unachieved goals heuristics is not good enough for itself, because it provides no guidance at all for achieving any of the goals, yet implicitly manages to yield a goal decomposition scheme that in many domains pays off, as indeed, most of the planning problems that are not inherently hard, are decomposable or nearly decomposable problems, where goals can be tackled one at a time, probably undoing slightly and reconstructing previously achieved goals [Korf, 1987, Simon, 1996]. In such cases, a best-first search guided by the number of unachieved subgoals heuristics, in which the additive heuristic is used as a tie-breaker, can do much better than either heuristic alone.

2.10   STRUCTURE, WIDTH, AND COMPLEXITY

There is a wide gap between the complexity of planning [Bylander, 1994], and the ability of current planners to solve most existing benchmarks in a few seconds. Work on *tractable planning* has been devoted to the identification of planning fragments that due to syntactic or structural restrictions can be solved in polynomial time; fragments that include for example problems with single atom preconditions and goals [Bylander, 1994, Jonsson and Bäckström, 1994]. On the other hand, work on *factored planning* has appealed instead to mappings of planning problems into Constraint Satisfaction Problems, and the notion of *width* over CSPs [Amir and Engelhardt, 2003, Brafman and Domshlak, 2006]. The width of a CSP is related to the number of variables that have to be collapsed to ensure that the induced graph underlying the CSP becomes a tree [Dechter, 2003, Freuder, 1982]. The complexity of a CSP is exponential in the problem width, and hence CSP-trees, for example, can be solved in linear time. A notion of width for classical planning using a form of Hamming distance was introduced by Chen and Giménez [2007], where the distance is set to the number of problem variables whose value needs to be changed in order to increase the number of goals achieved.

These proposals all identify planning fragments that can be solved efficiently, yet few of the existing benchmarks fall into these fragments even though they can be solved easily too. A related thread of research has aimed at understanding the performance of modern heuristic search planners by analyzing the characteristics of the *optimal delete-relaxation heuristic* h^+ that planners approximate for guiding the search for plans [Hoffmann, 2005, 2011]. For instance, the lack of local minima for h^+ implies that the search for plans (and hence the global minimum of h^+) can be achieved by local search, and this local search is tractable when the distance to the states that decrement h^+ is bounded by a constant. More recently, a novel *width* notion for classical planning has been introduced, along with a simple algorithm that solves problems in time and space exponential in their width [Lipovetzky and Geffner, 2012]. Basically, for a STRIPS planning problem $P = \langle F, I, O, G \rangle$, the authors define chains of tuples (sets) of atoms t_i, $t_0 \rightarrow t_1 \rightarrow \ldots \rightarrow t_n$ that obey two conditions: t_0 holds in the initial situation I, and for every optimal plan π that achieves jointly all the atoms in t_i from I, there is an action a such that π followed by a is an optimal plan for t_{i+1}. The sets of atoms t_i are like stepping stones in the construction of optimal plans for t_{i+1}, but stepping stones of a special type, where *all* optimal plans for t_i can be extended into optimal plans for t_{i+1}. A chain $t_0 \rightarrow t_1 \rightarrow \ldots \rightarrow t_n$ *implies* a formula W if all the optimal plans for t_n are also optimal plans for W. The size of the chain is the size of the largest set t_i in the chain, and the *width of the problem* P, $w(P)$, is the size of the min-size chain that implies the goal G of P. It is then shown that many of the existing benchmark domains have a bounded and low width when goals are restricted to *single atoms*, and an algorithm called *Iterated Width (IW)* search is presented that can solve planning problems in time exponential in their width. The algorithm is very simple; it is just a sequence of *pruned breadth-first searches* where a *novelty bound* used for pruning is increased from 0 until the problem is solved or the number of variables in the problem is exceeded. Basically, the *novelty* of a newly generated state s is defined as the size of the smallest tuple of atoms that is true in s and false in all the states generated in the search before s. Thus, for example, if s makes an atom p true that was false in all previously generated states, then the novelty of s is 1; if this is not true for any atom but is true for a pair of atoms (p, q), then the novelty of s is 2, and so on. If the problem has width $w(P)$, then it is shown that IW solves the problem in at most $w(P)$ iterations, in time and space exponential in $w(P)$. The blind-search algorithm IW, which does not look at the goal in any way, is shown to be quite effective in solving the standard domains when goals are single atoms, where it is competitive with the best heuristic-search planners. The blind-search algorithm IW is extended into another algorithm, Serialized IW (SIW), that uses IW for both decomposing a joint goal into atomic subgoals, and for solving each individual subgoal. Interestingly, while SIW is an incomplete blind-search algorithm, its performance over the benchmark domains is not far from the state of the art. While further work is required along this line, one tentative lesson that can be drawn from these results is that planners do well on most benchmarks because the benchmark domains are easy when containing single atomic goals, and because simple methods appear to work for decomposing joint goals into atomic goals. This account also suggests that problems with joint goals that are hard to decompose, and problems with atomic goals that have a high width, may constitute two sources of hard problems for classical planning.

CHAPTER 3

Classical Planning: Variations and Extensions

Most of the current state-of-the-art classical planners are based on the heuristic search formulation where plans are searched forward from the initial state using heuristics derived from the problem. This basic idea, however, has been extended in a number of ways, like in the use of structural information, also obtained automatically from the problem, in the form of "helpful actions" and "landmarks." In this chapter, we look at extensions and variations of the basic framework, at the heuristics developed for optimal planning, and at other computational approaches to classical and related forms of planning such as temporal and hierarchical task network planning.

3.1 RELAXED PLANS AND HELPFUL ACTIONS

The planner FF [Hoffmann and Nebel, 2001] that followed the first generation of heuristic search planners, used the relaxed plan heuristic inside a novel search architecture comprised of two phases: an incomplete but fast Enforced Hill Climbing (EHC) search, followed if needed by a complete but slower Greedy Best-First Search (GBFS). The gap in performance between FF and previous planners results mainly from the EHC search which can be particularly effective in problems where the branching factor is high. As explained in Section 2.8, the EHC search is a greedy search that uses a breadth-first lookahead where non-helpful actions are ignored and where the lookahead horizon is given implicitly by the first state that improves the value of the heuristic. It is the pruning of non-helpful actions that makes the EHC search incomplete. Recall that the helpful actions in a state s are defined as the actions applicable in the state s that add a goal or a precondition for an action in the relaxed plan $\pi_{FF}(s)$ that is not part of the state s. The reason that this lookahead is effective is because this notion of "helpful actions" often accounts for the actions that are most relevant to the goal, at least on domains that are not inherently hard. This, however, is an empirical observation; we don't understand yet why and when this is so.

A single state s in the EHC search where the action required is not helpful is sufficient to make the EHC search fail (assuming that this state can't be avoided in the way to the goal). The problem with this is that FF then switches to a complete GBFS where the information contained in the helpful actions is not used at all. The EHC will fail for example in a problem where an agent has to bring a suitcase to a destination, and there is a short route, say of 10 steps such that the agent cannot get past the fifth step with the suitcase, and a longer, disjoint route, say of 100 steps, with no such impediment. If the agent can drop the suitcase at any location and pick it up under the obvious preconditions, the relaxation will drive the EHC search along the short route, yet as long as the non-helpful action leading to the longer route is not taken in the initial state, the goal won't be achieved. The "helpful actions" are the actions that most directly drive the agent to the goal *in the relaxation*, yet these are not always the actions that serve best the agent in the original problem.

3.2   MULTI-QUEUE BEST-FIRST SEARCH

A key innovation in the planner Fast Downward (FD) that followed on FF [Helmert, 2006], was a way to use the information contained in the actions deemed as "helpful" in a complete search. For this, FD introduced a novel version of a Best-First Search algorithm that uses not just one OPEN list but multiple ones. The lists do not have to be disjoint and each one can be ordered by a different evaluation function. In particular, the complete algorithm that incorporates FF's heuristic along with FF's helpful actions in FD contains two OPEN lists, both using FF's heuristic as the evaluation function. Yet, the children nodes that result from the application of a helpful action are placed in the "helpful" OPEN list, while all the children, resulting either from helpful or non-helpful actions, are placed in the "non-helpful" OPEN list. The search algorithm then *alternates* between expansions of the best nodes according to the heuristic h_{FF} in the two OPEN lists [Röger and Helmert, 2010]. Of course, a node selected for expansion can be pruned when a duplicate of that node has been expanded already, and moreover, it makes sense to expand the "helpful" OPEN lists more often that the "non-helpful" list. Yet, as long as the selection between the two lists remains fair and one list is never left waiting forever, the algorithm remains complete and able to exploit information about helpful actions. Fast Downward complements this novel search architecture with a technique for dealing with the very large branching factors that result from not pruning the non-helpful actions as in the EHC search. The idea, called *delayed evaluation*, is to compute the heuristic of nodes only when they are *expanded* and not when they are *generated*. In the meantime, the heuristic of the node is set to the heuristic of its parent node (which must have been expanded). While this simplification leaves out useful information that could be derived, it avoids the overhead of evaluating many nodes that will never be expanded. Indeed, under delayed evaluation, the number of heuristic computations is equal to the number of nodes expanded rather than the possibly much larger number of nodes generated. Since the computation of the heuristic often represents the bulk of the computation in heuristic search planners, for problems with average branching factors of 100, for example, this technique can achieve time savings that can approach two orders of magnitude.

3.3   IMPLICIT SUBGOALS: LANDMARKS

We have seen in Section 2.9 that problem decomposition can pay off in some planning problems where a joint goal $G = \{G_1, \ldots, G_n\}$ can often be achieved by dealing with one subgoal G_i at a time, in an ordering that doesn't have to be fixed a priori. We have also seen that some of the standard planning heuristics such as h_{add} and h_{FF} do not deliver such a decomposition, while the simpler heuristic h_{ug} that counts the number of goals G_i not yet achieved, does.

The Multi-Queue Best-First search algorithm introduced in Fast Downward integrates multiple heuristics in a natural way by ordering different queues with different heuristics. The more recent planner LAMA [Richter and Westphal, 2010], which followed on Fast Downward, uses a similar search architecture for combining FF's heuristic with a more sophisticated version of the number of unachieved goals heuristic that counts not only the explicit top goals of the problem, but also *the implicit subgoals*. These implicit subgoals are called *landmarks*, which are defined as formulas that must achieved in the way to the goal by any plan that solves the problem [Hoffmann et al., 2004]. In particular, an *atomic landmark* is an atom that all plans for the problem must achieve at some point. LAMA uses both atomic landmarks and certain types of disjunctive landmarks. As an example, the atom $clear(A)$ is a landmark in any Blocks World problem where block A has to be placed onto a different block but its top is not clear. Similarly, if B is the block sitting on A, then the disjunction $ontable(B) \vee \bigvee_X on(B, X)$, where X ranges over the blocks X different than A, is also a landmark. While the problem of determining if a formula is a landmark is computationally intractable, there are very efficient algorithms for identifying some, although not necessarily all landmarks. In particular, one can identify landmarks in the delete-relaxation, as landmarks in the delete-relaxation are landmarks of the original problem too (this is because all plans are plans for the relaxation). An atom p will be a landmark of the problem when any of the delete-relaxation heuristics like h_{add}, h_{max}, or h_{FF} yield an infinite heuristic value in the problem that results from excluding the atom p from all action effects. Since all these heuristics yield an infinite value only when the delete-relaxation, and hence the original problem, is unsolvable, this means that if the problem is solvable at all, all the plans will have to contain an action that adds p. Since for this computation we are just interested in whether the value of the heuristic is infinite or not, for performance it pays off to set all action costs to zero when using either the additive or max heuristic for this purpose. While this computation has to be done $|F|$ times for identifying the atomic landmarks, where F is the set of atoms in the problem, this can be done just once as preprocessing. There are also algorithms that compute all atomic landmarks more efficiently in one pass, as well as algorithms that compute atomic and certain classes of disjunctive landmarks [Hoffmann et al., 2004, Keyder et al., 2010, Richter et al., 2008, Zhu and Givan, 2005].

The LAMA planner which has been the top performing planner in the last two International Planning Competitions [Coles et al., 2012, Helmert et al., 2008], uses the same Multi-Queue Best-First Search architecture as Fast Downward with two heuristics and four queues. The two heuristics are FF's and the number of *unachieved landmark heuristic*, which as we have seen, is an extension of the traditional but less informed number of unachieved top-goals heuristic. Two of the queues are then ordered by FF's heuristic, and two by the landmark heuristic. As in Fast Downward, one of the two queues for each heuristic is for the "helpful" children only; the other, for the non-helpful. The definition of "helpful" children is the standard one for the queue ordered by FF's heuristic (children of helpful actions); on the other hand, the helpful children in the landmark queue are defined in a different manner, as the children of actions that add an unachieved landmark, and when there are no such actions, as the actions that would be helpful in FF's sense for achieving the nearest unachieved landmark. These are somewhat arbitrary definitions that follow from performance considerations over the wide range of planning benchmarks, and there is not yet a good theory justifying these choices.

3.4 STATE-OF-THE-ART CLASSICAL PLANNERS

Table 3.1 shows the performance of a number of classical planners over benchmarks from past competitions, some of which feature hundreds of atoms and actions, and result in very long plans. We include the planners FF [Hoffmann and Nebel, 2001], Fast Downward [Helmert, 2006], LAMA [Richter and Westphal, 2010], Probe [Lipovetzky and Geffner, 2011], and BFS(f) [Lipovetzky and Geffner, 2012]. We have discussed most of these planners before. Probe is a GBFS planner guided by the additive heuristic that throws a carefully designed probe to the goal from each state that is expanded. The probe does not involve any search and either reaches the goal quickly or fails. In the first case, a plan is returned, in the second, the search continues. BFS(f) is also a GBFS planner but uses

Table 3.1: Some recent classical planners and their performance over competition benchmarks. Planners are FF, Fast Downward, Probe, LAMA, BFS(f). I is number of instances per domain, S is number of solved instances, Q and T are the average plan lengths and times in seconds computed over problems solved by all planners.

Domain	I	FF			FD			PROBE			LAMA'11			BFS(f)		
		S	Q	T	S	Q	T	S	Q	T	S	Q	T	S	Q	T
8puzzle	50	49	52.61	0.03	50	52.30	0.18	50	60.94	0.09	49	92.54	0.18	50	45.30	0.20
Barman	20	0	–	–	20	197.90	84.00	20	169.30	12.93	20	192.15	8.39	20	174.45	281.28
BlocksW	50	44	39.36	66.67	50	104.24	0.46	50	43.88	0.25	50	89.96	0.41	50	54.24	2.25
Cybersec	30	4	29.50	0.73	28	36.58	859.24	24	50.73	48.29	30	35.27	880.06	28	36.92	63.79
Depots	22	22	51.82	32.72	17	110.25	91.86	22	88.88	1.45	21	43.56	3.58	22	39.56	69.11
Driver	20	16	25.00	14.52	20	50.67	1.26	20	60.17	1.49	20	46.22	1.51	18	48.06	140.93
Elevators	30	30	85.73	1.00	30	92.57	3.20	30	107.97	26.66	30	97.07	4.69	30	129.13	93.88
Ferry	50	50	27.68	0.02	50	30.08	0.09	50	44.80	0.02	50	26.86	0.08	50	31.28	0.03
Floortile	20	5	44.20	134.29	3	39.00	6.91	5	40.50	106.97	5	40.00	8.94	7	36.50	4.15
Freecell	20	20	64.00	22.95	20	61.06	26.55	20	62.44	41.26	19	67.78	27.35	20	64.39	13.00
Grid	5	5	61.00	0.27	5	61.60	4.95	5	58.00	9.64	5	70.60	4.84	5	70.60	7.70
Gripper	50	50	76.00	0.03	50	152.62	0.17	50	152.66	0.06	50	92.76	0.15	50	152.66	0.38
Logistics	28	28	41.43	0.03	28	77.11	0.18	28	55.36	0.09	28	73.64	0.17	28	87.04	0.12
Miconic	50	50	30.38	0.03	50	39.80	0.07	50	44.80	0.01	50	31.02	0.06	50	34.46	0.01
Mprime	35	34	9.53	14.82	35	8.37	9.50	35	12.97	26.67	35	8.60	10.30	35	10.17	19.30
Mystery	30	18	6.61	0.24	19	6.86	1.87	25	7.71	1.08	22	7.29	1.70	27	7.07	0.93
NoMyst	20	4	19.75	0.23	6	22.40	1.96	5	23.20	2.73	11	23.00	1.77	19	22.60	0.78
OpenSt	30	30	155.67	6.86	30	130.11	5.97	30	134.14	64.55	30	130.18	3.49	29	125.89	129.06
OpenSt6	30	30	136.17	0.38	30	222.67	5.39	30	224.00	48.89	30	140.60	4.89	30	139.13	40.19
ParcPr	30	30	42.73	0.06	27	35.79	1.97	28	70.92	0.26	30	70.54	0.28	27	70.42	6.72
Parking	20	3	88.33	945.86	20	74.86	330.76	17	143.36	685.47	19	129.57	361.19	17	83.43	562.39
Pegsol	30	30	25.50	7.61	30	25.97	0.80	30	25.17	8.60	30	26.07	2.76	30	24.20	1.17
Pipes-N	50	35	34.34	12.77	44	75.50	7.94	45	46.73	3.18	44	54.41	11.11	47	58.39	35.97
Pipes-T	50	20	31.45	87.96	40	73.33	99.06	43	54.19	88.47	41	69.83	35.28	40	39.14	216.25
PSR-s	50	42	16.92	63.05	50	14.61	0.27	50	17.20	0.07	50	14.65	0.31	48	18.14	2.57
Rovers	40	40	100.47	31.78	40	153.18	13.69	40	131.20	24.19	40	108.53	17.90	40	126.30	44.20
Satellite	20	20	37.75	0.10	20	40.90	0.78	20	37.05	0.84	20	42.05	0.78	20	36.05	1.26
Scan	30	30	31.87	70.74	28	30.04	7.30	28	25.15	5.59	28	28.04	8.14	27	29.37	7.40
Sokoban	30	26	213.38	26.61	28	204.14	12.44	25	231.52	39.63	28	231.81	184.38	23	218.52	125.12
Storage	30	18	16.28	39.17	20	17.72	3.20	21	14.56	0.07	18	24.56	8.15	20	20.94	4.34
Tidybot	20	15	63.20	9.78	15	66.00	338.14	19	52.67	33.50	16	62.60	102.52	18	63.27	207.85
Tpp	30	28	122.29	53.23	30	127.93	16.95	30	152.53	60.95	30	205.37	18.72	30	110.13	126.03
Transport	30	29	117.41	167.10	30	97.57	12.75	30	125.63	38.87	30	215.90	76.18	30	97.57	46.64
Trucks	30	11	27.09	3.84	17	26.00	0.65	8	26.75	113.54	16	24.75	0.53	15	26.50	8.59
Visitall	20	6	450.67	38.22	7	3583.86	166.35	19	411.71	9.02	20	468.00	4.68	20	339.00	4.58
WoodW	30	17	32.35	0.22	30	57.13	18.40	30	41.13	15.93	30	79.20	12.45	30	41.13	19.12
Zeno	20	20	30.60	0.17	20	37.45	2.68	20	44.90	6.18	20	35.80	4.28	20	37.70	77.56
Summary	1150	909	67.75	51.50	1037	168.60	57.78	1052	83.64	41.28	1065	86.51	48.98	1070	74.32	63.91

an evaluation function based on width-considerations (Section 2.10) along with tie breakers based on the additive and number-of-unachieved-landmark heuristics. The scalability of planners has improved considerably over the last 15 years with the best planners using and extending the ideas of previous planners, such as heuristic functions, helpful actions, and landmarks. As a reference, a baseline planner such as HSP, based solely on a Greedy Best-First Search guided by the additive heuristic, solves 789 of the problems, while LAMA, the winner of the last two competitions solves 1,065 problems out of a total of 1,150. In the Table, I stands for the number of instances per domain, while S, Q, and T stand for the number of instances solved, and the average plan lengths and times in seconds. The experiments were conducted on a dual-core CPU running at 2.33 GHz and with two GB of RAM, with processes timing out after two hours. All of these domains and planners, including their sources, are available on the Internet.

3.5   OPTIMAL PLANNING AND ADMISSIBLE HEURISTICS

Optimal planners ensure the optimality of the plans found by using admissible heuristics (lower bounds) in the context of search algorithms like A* or IDA*. Most admissible heuristics developed for planning are based either on the *delete-relaxation* like the heuristic h_{max} [Bonet and Geffner, 2001], on a notion of *critical paths* like the heuristics h^m [Haslum and Geffner, 2000], on *abstractions* where certain atoms are dropped from the problem, like pattern-database heuristics [Edelkamp, 2001], or on *landmarks*, like the LA and LM-Cut heuristics [Helmert and Domshlak, 2009, Karpas and Domshlak, 2009]. Many of these heuristics are general templates that leave some choices open, and some dominance relations among these heuristics have been established as well [Helmert and Domshlak, 2009].

The heuristics h^m, where m is a positive integer, are based on the assumption that the estimated cost $h^m(C, s)$ of achieving a (conjunctive) set of C atoms from a state s is given by the estimated cost $h^m(C', s)$ of achieving the most costly subset C' of at most m atoms in C. Mathematically, the estimate $h^m(C, s)$ is defined inductively as:

$$h^m(C, s) = \begin{cases} 0 & \text{if } C \subseteq s, \\ \min_{a \in R(C)} [c(a) + h^m(Reg(a, C), s)] & \text{if } C \not\subseteq s \text{ and } |C| \leq m, \\ \max\{h^m(C', s) : C' \subseteq C, |C'| \leq m\} & \text{otherwise,} \end{cases} \quad (3.1)$$

where $Reg(a, C)$ stands for the regression of the set of atoms C through the action a, i.e., $Reg(a, C) = (C \setminus Add(a)) \cup Prec(a)$, and $R(C)$ stands for set of actions a in the problem that add some atom in C and delete none. The approximation captured by this definition follows from setting the estimated cost $h^m(C, s)$ of achieving sets C of more than m atoms, to the cost of achieving the most costly subset C' of C of at most m atoms. The h^m heuristic for state s, $h^m(s)$, is $h^m(s) = h^m(G, s)$, where G is the goal of the problem. For $m = 1$, it is easy to show that h^m is equal to h_{max}, while for a sufficiently large value of m that is less than or equal to the total number of variables in the problem, h^m is equal to the optimal heuristic h^*.

Pattern database (PDB) heuristics provide a generalization of the PDB heuristics developed for domain-specific heuristic search [Culberson and Schaeffer, 1998]. A PDB heuristic is a *lookup table* that stores exact optimal distances for an abstraction of the problem computed by a regression search from the goal. The abstraction is obtained by dropping a sufficient number of atoms from the problem so that the number of reachable states in the reduced problem fits in memory. An atom is removed from a problem $P = \langle F, I, O, G \rangle$ by removing it from F, I, O, and G; i.e., the atom is removed from precondition, delete, and add lists, from the goal and the initial situation, and from the set of problem atoms. If s is a state over the original problem, and A is the set of atoms retained in the reduced problem, the heuristic $h(s)$ is set to the optimal heuristic $h^*(s')$ over the reduced problem where s' is the state $s' = s \cap A$. A key challenge in the design of PDBs is deciding which atoms to abstract away from the problem [Haslum et al., 2007]. Two recent variations on the PDB idea for planning are the merge-and-shrink heuristics [Helmert et al., 2007] and structural pattern heuristics [Katz and Domshlak, 2008b].

Multiple admissible heuristics $h_1, \ldots, h_n$ can be combined into a potentially more informed admissible heuristic by taking their pointwise maximum as $h(s) = \max\{h_1(s), \ldots, h_n(s)\}$, or by partitioning the action costs [Haslum et al., 2007, Katz and Domshlak, 2008a]. A *cost partitioning* Π of problem P with cost function $c(\cdot)$ is a collection $P_1, \ldots, P_n$ of problems identical to P except on their cost functions $c_1, \ldots, c_n$ that must be non-negative and satisfy $\sum_{i=1,n} c_i(a) \leq c(a)$ for every action a in P. Then, if $h_1, \ldots, h_n$ are (arbitrary) admissible heuristics for the problems $P_1, \ldots, P_n$ respectively, the additive heuristics $h = h_1 + \cdots + h_n$ is an admissible heuristic for P. One can indeed improve a base heuristic by doing a cost partitioning that applies the same base heuristic to every problem in the partition; the difficulty is in the choice of the cost functions c_i, $1 \leq i \leq n$, and the number of partitions.

The LM-Cut heuristic [Helmert and Domshlak, 2009] is a powerful admissible heuristic that can be thought as either a landmark heuristic or a cost partitioning heuristic based on h_{max}. For determining the LM-Cut value of a state, a sequence $L_1, \ldots, L_m$ of (disjunctive but not necessarily disjoint) *action landmarks* are computed together with cost functions $c_1, \ldots, c_m$ providing a cost partitioning. Like (atomic) landmarks, a disjunctive action landmark L for state s is a set of actions such that every plan from the state s must contain one of the actions in the set. In cases when the landmarks computed by LM-Cut are pairwise disjoint, the cost function c_1 in the partition defined by LM-Cut assigns costs $c_1(a) = c(a)$ to every action $a \in L_1$ and $c_1(a) = 0$ to $a \notin L_1$, costs $c_2(a) = c(a)$ for every action $a \in L_2 \setminus L_1$, and $c_2(a) = 0$ for $a \notin L_2$, and so on, while the heuristic values in such cases become $h_1 = \min_{a \in L_1} c_1(a)$, $h_2 = \min_{a \in L_2} c_2(a)$, and so on. The value of the LM-Cut heuristic at state s is the sum $h_1 + h_2 + \cdots + h_m$ as in any cost partitioning scheme, which is guaranteed to be admissible. The LM-Cut heuristic can be improved by computing and considering more than one landmark at a time, exploiting a connection between action landmarks and hitting sets [Bonet and Helmert, 2010].

3.6 BRANCHING SCHEMES AND PROBLEM SPACES

We have reduced the search for plans for a STRIPS planning problem $P = \langle F, I, O, G \rangle$ to the search for paths in the directed graph associated with the state space $S(P)$, where the root node represents the initial state I, the target nodes represent the states that include the goal G, and the edges express the state transitions that are possible given the actions in O. Yet, the search for plans can be formulated in many other ways, some of which will be explored over the next few sections. As we will see, the search for plans can also be formulated as a path-finding problem over a different graph, so that plans are searched backward from the goal rather than forward from the initial state. We will also see formulations where plans are not constructed directionally either from the initial situation or from the goal. In all cases, however, the search will involve the traversal of a graph, along with heuristics or inference procedures for selecting or pruning branches. Branches will represent partial plans: either partial plans that have to be refined further, partial plans that are complete and hence encode solutions, or partial plans that can't be refined into solutions. In the formulation we have considered so far, branches in the graph represent *plan prefixes*, and the nodes at the end of these branches represent the resulting states. In other formulations, branches represent *plan suffixes*, *partially ordered plans*, or simply *commitments* about actions and propositions that are true or false at different time steps. While the formulation considered so far is the most natural, and the one that underlies the best current planners, the other formulations are also useful, and may in fact be superior in slightly different settings, as in *temporal planning*, where actions have different durations and may be executed concurrently, and plans that minimize total duration (makespan) are sought. The graph that results from these decisions, and in particular, the way in which children nodes are generated from parent nodes in the graph, are sometimes referred to as the *branching scheme*, the *problem space*, or the *search graph*. While in the standard problem space for planning, the branching factor (number of children) is given by the number of applicable actions, we will see other problem spaces for planning whose branching factor is always 2, regardless of the number of actions or atoms in the problem. As an illustration of this, consider the famous Travelling Salesman Problem (TSP), where a minimum cost tour that visits every node in a graph once is sought [Lawler *et al.*, 1985]. The most natural way for solving the problem is by starting in an arbitrary TSP node n, and by defining the children of the node in the search graph as the TSP nodes that can be reached from n in one step. Branches in this search graph would thus stand for tour prefixes. In principle, if there are n nodes in the TSP graph, this means up to $n - 1$ children per node in the search graph, and hence a large branching factor if the TSP graph is large and dense. A different branching scheme for solving the TSP is to pick one edge (i, j) from the TSP graph, while considering two options: that the edge will be part of the optimal solution, or that the edge will not be part of the optimal solution. The resulting search graph then becomes a tree with a branching factor of 2, where the branches are no longer *tour prefixes* but *tour commitments*, namely, that certain edges in the TSP graph have to be part of the final tour and that other edges don't. Interestingly, in combination with good admissible heuristics for the TSP like the assignment problem heuristic, this less direct form of branching, scales up much better for large TSPs than the most straightforward approach of starting in one city and building the tour forward from it [Lawler *et al.*, 1985]. It is crucial though that the problem space and the heuristics or inference used for guidance or pruning are well matched to each other. Heuristics such as h_{FF} and h_{add} work well in the forward search for plans because the initial state of the problem can be progressed through the plan prefix (by applying the actions in sequence), and the heuristic for the plan prefix can be computed from the state that results. This property extends also to *regression planners* that search for plans backward, where a suitable state, obtained by *regressing the goal through a plan suffix*, is used for computing the heuristic. For this reason, planners that search for plans forward or backward are often called *state-space planners*. Computing heuristics for planners that search for plans in other ways is more difficult, yet powerful propagation and pruning criteria have been developed for SAT and CSP formulations.

3.7   REGRESSION PLANNING

Planners can search for plans backward from the goal by applying the actions in reverse. Basically, if a set of atoms C is to be achieved, the only actions that could be "last" in a minimal plan for achieving

C are actions that add some atom in C and do not delete any. If a is one such action, the set of atoms C' that have to be true right before the application of a in order for C be true right after, are the action preconditions of a along with the atoms in C that are not added by a. The set of atoms C' is said to be then *the regression of C through the action a*. Starting this "subgoaling" process with the top goal G of the problem, we obtain a state space that is called the *regression state model* associated with the STRIPS planning problem $P = \langle F, I, O, G \rangle$, to distinguish it from the forward or *progression state model* $\mathcal{S}(P)$ [Bonet and Geffner, 1999, Nilsson, 1980, Weld, 1994]. In this regression space $\mathcal{R}(P)$:

- the states s are sets of atoms from F,

- the initial state s_0 is G,

- the goal states are the states s for which $s \subseteq I$,

- the set of actions $A(s)$ applicable in s are the actions $a \in O$ that are *relevant* and *consistent*, namely, for which $Add(a) \cap s \neq \emptyset$ and $Del(a) \cap s = \emptyset$,

- the state $s' = f(a, s)$ that follows the application of action $a \in A(s)$ in s is $s' = (s \setminus Add(a)) \cup Prec(a)$, and

- the action costs are $c(a, s)$.

The solution of this state space is like the solution of any state model, an applicable action sequence mapping the initial state into a goal state. Yet, notice that in the regression space $\mathcal{R}(P)$, the initial state is the goal of P, and the goal states are the states that must be true in the initial situation of P. Moreover, while the states s in the regression space $\mathcal{R}(P)$ are defined syntactically as in the progression space $\mathcal{S}(P)$ by *sets of atoms*, the meaning of these sets is very different in the two cases: states represent *complete truth-assignments* in the progression space but *partial truth-assignments* in the regression space. In particular, in the initial state $s_0 = I$ of the *progression space $\mathcal{S}(P)$*, every atom that is not in I is false, while in the initial state $s_0 = G$ of the *regression space*, this is not true; indeed, the goal G can be the single atom $on(A, B)$ in a blocks world problem with three blocks A, B, and C, and there is no reachable state of the problem where $on(A, B)$ is true and all other atoms are false (blocks B and C must be somewhere!). It can be shown that every state s in the regression space $\mathcal{R}(P)$ stands indeed for a collection of states s' in the progression space $\mathcal{S}(P)$, namely all the states s' in the progression space that include s. Alternatively, the states s in the regression can be thought of as *goals* and *subgoals* to be achieved, with an action applied in reverse, mapping one goal into another one.

The regression space $\mathcal{R}(P)$ is sound and complete in the sense that the solutions to $\mathcal{R}(P)$ encode the solutions to the problem P but in reverse. One potential advantage of searching for plans backward in $\mathcal{R}(P)$ as opposed to forward in $\mathcal{S}(P)$ is that it is possible to avoid the computation of the heuristic from scratch in every state. This can represent up to 80% of the total time that heuristic search planners take in solving a problem. In the regression search it is indeed possible to perform the bulk of this computation just once. For example, if $h(p; s_0)$ is the estimated cost of achieving the atom p from the initial problem state s_0 according to the additive heuristic, the estimated cost $h(s')$ from a state s' to the goal I *in the regression search* can be set up to the sum $h(s') = \sum_{p \in s'} h(p; s_0)$, where the elements of the sum need to be computed just once from s_0 and used then to determine the heuristic $h(s')$ of any state s' in the regression space.

A potential problem in the regression search is that while the regression is sound and complete, it may contain many more dead-ends than the progression search; i.e., the regression can generate collection of atoms that cannot be achieved jointly by any plan from the initial situation, and hence that will not lead to any solution [Bonet and Geffner, 1999]. In addition, it is not clear that the heuristics as defined above are as informative in the backward search as in the forward search. None of these issues have been studied throughly, though, yet the fact is that there are no regression-based planners these days that can compete with the best progression-based planners examined above.

3.8 PLANNING AS SAT AND CONSTRAINT SATISFACTION

SAT is the problem of determining whether a formula in Conjunctive Normal Form (CNF) is satisfiable. A formula in CNF can be regarded as a set of disjunctions of literals where a literal is an atom or its negation. Such disjunctions are called *clauses*. For instance, the formula $(p \vee \neg q) \wedge (\neg p \vee q)$ is in CNF, $p \vee \neg q$ and $\neg p \vee q$ are its clauses, and the formula is satisfiable as the assignment $p \mapsto true$, $q \mapsto true$, for example, makes both clauses true. SAT is an NP-Complete problem [Sipser, 2006], which in practice means that all complete algorithms for SAT will run in exponential time in the worst case. Still, very large SAT instances can be solved nowadays due to very effective inference techniques, such as Unit Propagation and Conflict-driven Learning [Biere et al., 2012]. The problem of classical planning can be expressed as a SAT problem provided that a planning horizon is given. The SAT approach to planning has been introduced and shown to be effective by Kautz and Selman in the mid 90s [Kautz and Selman, 1992, 1996, 1999]. The basic idea is very simple. For a STRIPS problem $P = \langle F, I, O, G \rangle$ and a planning horizon n, a CNF formula $C(P, n)$ is produced that is fed into a general SAT solver. The CNF formula $C(P, n)$ includes propositions $p_0, p_1, \ldots, p_n$ for each atom $p \in F$ and propositions $a_0, a_1, \ldots, a_{n-1}$ for each action $a \in O$. The formula $C(P, n)$ is such that $C(P, n)$ is satisfiable iff there is a plan of at most n steps that solves the problem P. In such a case, the plan can be extracted from the actions that are true in the satisfying assignment.

The clauses in the formula $C(P, n)$ encoding the planning problem $P = \langle F, I, O, G \rangle$ with horizon n are given in Figure 3.1. It can be shown that if $C(P, n)$ is not satisfiable, then there is no plan of length n solving the problem P, while if $C(P, n)$ is satisfiable, the plan given by the action that is true at time 0, followed by the action that is true at time 1, and so on, with truth evaluated in the satisfying assignment, is a plan that solves P. Moreover for any plan that solves P, there is an assignment satisfying the formula $C(P, n)$ that makes the plan true. Since the horizon n required to find a plan is not known a priori, the SAT approach to planning increases the horizon one by one from $n = 0$ until a satisfying assignment for $C(P, n)$ is found.

The approach, as described, works remarkably well, although it doesn't scale up as well as the best heuristic search planners. In part, this is to be expected, as in this formulation the first plan found is an optimal plan (assuming actions costs are uniform). In order for the SAT approach to scale up, a number of variations have been introduced. One is to allow certain *sets of actions* to be done in *parallel*, in particular, the sets of actions that if applicable in a given situation, remain applicable and yield the same overall effect regardless of how they are ordered. A sufficient condition for this is that no action in the set deletes a precondition or add effect of any other action in the set. Actions that do not comply with this condition are called *mutex*, as they are regarded as mutually exclusive at any one time step [Blum and Furst, 1995]. Other improvements involve the introduction of a NO-OP action for each atom p in the problem, with precondition and effect p, which helps to simplify the

Init: p_0 for each $p \in I$, $\neg q_0$ for each $q \in F$ such that $q \notin I$.

Goal: p_n for each $p \in G$.

Actions: For $i = 0, 1, \ldots, n - 1$ and each action $a \in O$:
  $a_i \supset p_i$ for each $p \in Pre(a)$
  $a_i \supset p_{i+1}$ for each $p \in Add(a)$
  $a_i \supset \neg p_{i+1}$ for each $p \in Del(a)$

Persistence: For $i = 0, 1, \ldots, n - 1$ and each fluent $p \in F$, where $O(p^+)$ and $O(p^-)$ are the actions that add and delete p respectively,

$$p_i \wedge \bigwedge_{a \in O(p^-)} \neg a_i \supset p_{i+1}$$
$$\neg p_i \wedge \bigwedge_{a \in O(p^+)} \neg a_i \supset \neg p_{i+1}$$

Seriality: For each $a, a' \in O$ such that $a \neq a'$, $\neg(a_i \wedge a'_i)$ for $i = 1, \ldots, n - 1$.

Figure 3.1: The subformulas that make up the CNF formula $C(P, n)$ encoding a planning problem $P = \langle F, I, O, G \rangle$ with horizon n. Fluents p and actions a are tagged with time indices i. The subformulas can be easily converted into clauses.

clauses encoding the persistence axioms, and the use of lower bounds for initializing the planning horizon [Kautz and Selman, 1999]. More recently, Jussi Rintanen has introduced other refinements that improve the performance of SAT-based planners further, including the use of a special heuristic for variable selection in the otherwise generic SAT solver, an improved search for an adequate planning horizon, and better memory management for dealing with the millions of clauses that often result from planning encodings in CNF [Rintanen, 2012]. While SAT-based planners do not yet scale up as well as the best heuristic-search planners, the gap has narrowed down considerably. Moreover, it is well known that on domains that are inherently difficult, SAT approaches can do much better than optimal and non-optimal heuristic search planners [Hoffmann et al., 2007].

Constraint Satisfaction Problem (CSP) provide a generalization of SAT where the variables are not restricted to be boolean, and constraints are not restricted to be clauses [Dechter, 2003]. General CSP solvers can deal with multivalued variables over arbitrary constraints. In the same way that classical planning problems with a fixed horizon can be cast as SAT problems, they can also be cast as CSP problems [Do and Kambhampati, 2000]. In spite and perhaps because of the additional expressive power afforded by the CSP formulation, CSP-based planners have not been able to keep up with SAT-based planners, that are based on a more restricted task (SAT) over which the technology has moved faster.

3.9 PARTIAL-ORDER CAUSAL LINK PLANNING

Branching schemes have also been developed for searching for plans neither forward from the initial state or backward from the goal, but by extending a set of actions that are partially ordered. The resulting planners are known as *partial-order planners* [Ghallab et al., 2004, Weld, 1994]. We follow below the formulation of partial order planning known as *partial-order causal-link* or POCL planning, where structures known as *causal links* are used in the plan representation to keep track of temporal precedences and constraints [McAllester and Rosenblitt, 1991]. A *partial plan* σ in POCL planning corresponds to a set of commitments represented by a tuple $\sigma = \langle Steps, Ord, CL, Open \rangle$, where $Steps$ is the set of actions in the partial plan σ, Ord is a set of precedence constraints on $Steps$, CL is a set of causal links, and $Open$ is a set of open preconditions. A precedence constraint $a \prec a'$ states that action a precedes action a' in the partial plan, a *causal link* $a[p]a'$ states that action a supports the precondition p of action a' in the partial plan, and an *open precondition* $[p]a$ states that the precondition p of the action a in the partial plan is open, meaning that it is not yet supported by any action. The initial node of the search, σ_0, is given by the tuple $\langle \{Start, End\}, \{Start \prec End\}, \emptyset, \{[G_1]End, \ldots, [G_m]End\} \rangle$ where $G_1, G_2, \ldots, G_m$ are the top level goals of the problem. Here $Start$ and End are two dummy actions used for encoding the initial situation and goals: the action $Start$ must precede all actions and is assumed to add all the atoms in the initial situation of the problem, and the action End must be the last action, whose preconditions are the goals of the problem and whose effect is the dummy target goal G.

Branching in POCL planning proceeds by picking a "flaw" in a non-terminal node (partial plan) σ and applying the possible *repairs* [Kambhampati et al., 1995, Weld, 1994]. Flaws are of two types. *Open precondition flaws* $[p]a$ in σ are solved by selecting an action a' that supports p and adding the causal link $a'[p]a$ to CL and the precedence constraint $a' \prec a$ to Ord (a' should also be added to $Steps$ if $a' \notin Steps$). Similarly, *threats*—which refer to situations in which an action $a \in Steps$ deletes the condition p in a causal link $a_1[p]a_2$ in CL with the ordering $a_1 \prec a' \prec a_2$ consistent with Ord—are solved by placing one of the precedence constraints $a' \prec a_1$ or $a_2 \prec a'$ in Ord. A node is terminal if it is inconsistent (i.e., the ordering Ord is inconsistent or contains flaws that cannot be fixed) or is a *goal* (is consistent and contains no flaws). The goal nodes represent partial plans that are complete and solve the problem. A plan can then be obtained from such nodes by any total ordering of $Steps$ that respects the partial order Ord; an operation that can be done in polynomial time [Dechter et al., 1991].

POCL planning is a clever branching scheme for organizing the search for plans, which is sound and complete. Up to the mid 90s, it represented the main computational approach in planning, but with the advent of Graphplan, SAT, and heuristic-based planners, it lost appeal as it could not scale up as well. One of the main problems is the lack of good heuristics for guiding or pruning the POCL search. POCL planning, however, remains a convenient scheme for some expressive forms of planning, including planning with time, resources, and concurrency [Smith et al., 2000], and several attempts have been made to make its search more informed [Nguyen and Kambhampati, 2001, Vidal and Geffner, 2006].

3.10   COST, METRIC, AND TEMPORAL PLANNING

Optimal and non-optimal planning in the presence of non-uniform action costs $c(a)$ is direct in the heuristic search approach to planning, where most heuristic estimators can be easily extended to take action costs into account. In SAT or CSP approaches, the handling of such costs is less direct. In all cases, one important challenge that has been left unaddressed is the handling of costs $c(a, s)$ that depend on both the action and the state where it is applied, or costs $c(s)$ that depend solely on the state. One could express, for example, that states s where two atoms p and q are true, are to be avoided, by assigning a high cost $c(s)$ to actions applied in such states. Unfortunately, no informative heuristics have been formulated so far for taking such state-dependent costs into account.

Numeric or metric planning refers to planning in the presence of numeric variables that can potentially take an infinite number of values, such as all integer, rational, or real numbers, or whose range cannot be bounded a priori. If this range is finite, the problem can be transformed into a planning problem over a finite collection of multivalued or boolean variables, even if this transformation is not always convenient computationally.

In the presence of such numerical variables X, the states are defined in the usual way as assignment of values x to variables X, and the initial situation fully defines one such state. Goals and preconditions, however, can then include atoms like $X = Y$, $X > Y$, $X \geq Y$, and their negations, where Y is a value or another variable, and the effect of these actions can involve expressions of the form $X := f(X_1, \ldots, X_n)$ where the new value of X is set as a function of the current value of a subset of variables X_i that may include the variable X itself. In common applications of metric-planning, numeric variables refer to resources: money, fuel, time, space, etc.

The semantics of metric planning problems is direct, as they easily map into classical planning models where a target state is to be achieved from a given initial state by applying deterministic actions. The sole difference with the standard classical planning model is that the number of states in metric planning may be infinite. While in the case of non-integer variables this may present some subtleties (e.g., no finite plan can achieve $X = 0$ if the only available action changes X to $X := X/2$), the challenge that has received the most attention in metric planning is the search for finite plans (when they exist), and in particular, the formulation and use of heuristics that take numeric variables into account. Metric-FF was one of the first modern planners to tackle this problem by introducing a polynomial relaxation and heuristic that rather than ignoring "negative" effects, as in the delete-relaxation, it ignores either increasing or decreasing effects [Hoffmann, 2003]. The relaxation yields lower and upper bounds for the numeric variables so that values within these bounds are assumed to be all achievable. Thus, in particular, a relaxed plan that makes the upper bound of X higher than the lower bound of Y is assumed to be also a relaxed plan for an atom like $X > Y$. The relaxation provides useful guidance in many problems, although there are obvious limitations such as potentially regarding an atom like $X > X$ as achievable. More recent work addressing metric planning can be found in [Coles et al., 2009, Do and Kambhampati, 2001, Edelkamp, 2006, Gerevini et al., 2008, van den Briel and Kambhampati, 2005].

Finally, temporal planning refers to planning in the presence of actions with durations, that under certain conditions can be executed concurrently. Normally, the objective in temporal planning is to achieve the goal as early as possible, often referred to as the minimization of the plan makespan. From a computational point of view, temporal planning raises two challenges. The first has to do with the problem space: the branching factor that results from the explicit consideration of *sets of parallel actions* in a forward search may be just too large. In general, it is thought that the problem space associated with POCL planning, where actions in the plan are partially ordered is then more convenient [Smith et al., 2000]. This, however, raises the second challenge: the control of the search in temporal POCL planning. CPT is a temporal planner that optimizes makespan by means of a constraint programming formulation of POCL planning where partial plans that cannot be refined into complete plans within a given makespan are detected and pruned early in the search by a form of constraint propagation [Vidal and Geffner, 2006]. A common strategy when formal guarantees on the makespan are not required is to deal with temporal problems as if they were standard sequential problems with action costs set to action durations. Then different approaches can be used to parallelize the resulting plans for reducing the makespan. This approach, however, is not universal, and can be used only when temporal plans, accommodating concurrent actions, can be serialized. It can be shown that this is true, for example, when the only sets of actions that can be done in parallel, i.e., that may overlap in time, are commutative actions, like non-mutex actions that do not delete a precondition or effect of another action [Blum and Furst, 1995, Smith and Weld, 1999]. Often, however, there are parallel plans that cannot be serialized in this way, as when someone must light a match for inserting a key in a door. Interestingly, the current version of the planning language standard, PDDL 2.1 [Fox and Long, 2003], can express temporal problems of this type, which involve what is called *required concurrency* [Cushing et al., 2007]. While most existing temporal planners, whether optimal or not, do not handle this type of concurrency, the extensions required for handling it may not be that complex in some formulations. In a constraint-based temporal planner like CPT, this may just require the ability to express and deal with simple constraints on actions, such as that if a plan includes one action a at time t, then it must include another action (event) b at time $t + \Delta$, like "if light is turned on now, it'll turn off in 30 seconds." Such action constraints create a new type of "flaws" in partial plans that must be fixed too, e.g., by including b in the partial plan at time $t + \Delta$ when a is in the plan at time t. The expressive power afforded by such changes and the necessary ways for dealing with them, however, have not been fully addressed yet.

3.11   HIERARCHICAL TASK NETWORKS

In the forms of planning considered so far, no information is given about which actions to apply or which subgoal to pursue; rather, actions are characterized in terms of their pre and postconditions, and the choice and ordering of the actions for solving a problem is computed automatically. HTN planning, where HTN stands for Hierarchical Task Networks, provide a completely different way of constructing plans [Erol et al., 1994, Ghallab et al., 2004]. In HTN planning, plans are not obtained from a model that describes how the actions change the world, rather actions, called tasks, are described at several levels, with tasks at one level decomposing into tasks at a lower level, with some tasks, called the primitive tasks, standing for real executable actions that do not decompose further. For example, the abstract task of taking a taxi may be decomposed into the tasks of getting to the street, stopping a taxi, getting on board, and so on. Similarly, the task of getting on the taxi can be decomposed into the tasks of opening the door, entering the taxi, and closing the door. Some tasks, like getting to the airport, may admit multiple decompositions called *methods*, one of which may involve the tasks of taking a taxi to the train station, and then a train from the station to the airport. In this case, the tasks inside the methods are constrained to be one after the other. Other types of restrictions may relate the tasks in a method as well. While the objective in classical planning is to find an action sequence that maps the initial situation into a goal state, in HTN planning, the objective is to find a decomposition that results in a consistent network of primitive tasks. Classical planning is model-based because it's based on a model of the actions, the initial situation, and goals, from which the plan is derived and with respect to which the computed plan can be proved correct. HTN planning is not model-based in this sense, and indeed, in HTN planning there is no clear separation between the problem that is being solved and the strategies being used for solving it. Actually, HTNs are most commonly used for encoding *solution strategies*. From a theoretical point of view, this is not good enough, as there is then no assurance that the planning strategy encoded by the HTN leads to plans that are correct.[1] Yet, from a practical point of view, this may be a feature rather than a bug: in many applications, humans feel more comfortable describing the solving strategies for the domain than the domains themselves, and place more trust on such strategies than on plans found by domain-independent planners. This may explain why HTN planners are more common in applications than domain-independent planners. This, however, may change, as better ways are found for integrating strategy and domain descriptions, and for coming up automatically with general and transparent strategies.

---

[1]Indeed, in some of the *knowledge-based* planning competitions held so far, teams were given the planning domains in advance, and they were free to determine and encode the strategies for solving them. It was then found that plans obtained from these strategies were not always correct. This is because the HTN encodings were not derived from the domain descriptions but were written by hand.

CHAPTER 4

Beyond Classical Planning: Transformations

We have considered models of planning where a goal is to be achieved by performing actions that are deterministic given an initial situation that is fully known. Often, however, planning problems exhibit features that do not fit into this format, features such as goals that are desirable but which are not to be achieved at any cost (soft goals), goals that refer not only to end states but to the intermediate states as well (temporally extended goals), or initial situations that are not fully known (conformant planning). In this chapter, rather than reviewing more powerful algorithms for dealing with such features, we illustrate how features such as these can be handled by off-the-shelf classical planners through suitable transformations that can be performed automatically. Similar transformations will also be introduced for dealing with a different task, plan recognition, where a probability distribution over the possible goals of the agent is to be inferred from partial observations of the agent behavior.

4.1 SOFT GOALS AND REWARDS

Soft goals are used to express desirable outcomes that unlike standard hard goals are subject to a cost-utility tradeoff [Sanchez and Kambhampati, 2005, Smith, 2004]. We consider a simple STRIPS planning setting where problems $P = \langle F, I, O, G \rangle$ are extended with information about positive *action costs* $c(a)$ for every action $a \in O$, and non-negative *rewards* or *utilities* $u(p)$ for every atom $p \in F$. The soft goals of the problem are the problem atoms with positive utility. It is possible to associate utilities with more complex logical formulas, like disjunctions of atoms or negated literals, yet standard methods can be used to introduce atoms for representing such formulas [Gazen and Knoblock, 1997].

In the presence of soft goals, the target plans π are the ones that maximize the utility measure or *net-benefit* given by the difference between the total utility obtained by the plan and its cost:

$$u(\pi) = \sum_{p:\pi \models p} u(p) - c(\pi) \qquad (4.1)$$

where $c(\pi)$ is given by the sum of the action costs in π, and $\pi \models p$ expresses that p is true in the state that results from applying the action sequence π to the initial problem state.

A plan π for a problem with soft goals is optimal when no other plan π' has utility $u(\pi')$ higher than $u(\pi)$. The utility of an optimal plan for a problem with no hard goals is never negative as the empty plan has non-negative utility and zero cost. The International Planning Competition held in 2008 featured a *Net-Benefit Optimal* track where the objective was to find optimal plans with respect to Eq. 4.1 [Helmert et al., 2008]. Soft goal or net-benefit planning appears to be very different than classical planning as it involves two interrelated problems: deciding which soft goals to adopt, and deciding on the plan for achieving them. Indeed, most of the entries in the competition developed native planners for solving these two problems. More recently, however, it has been shown that problems P with soft goals can be compiled into *equivalent problems P' without soft goals* that can then be solved by classical planners able to handle action costs $c(a)$ only [Keyder and Geffner, 2009]. The plans for P and P' are the same, except for the presence of dummy actions, and the utilities of the plans for P are inversely related to the cost of the plans for P'. Thus, optimal cost-based planners for P' yield optimal net-benefit plans for P, while satisficing cost-based planners for P', that scale up better, yield satisficing net-benefit plans for P.

The idea of the transformation from the problem P with soft goals into the equivalent problem P' with hard goals only is very simple. For soft goals p associated with individual atoms, one just needs to add new atoms p' that are made into *hard goals* in P', that are achievable in one of two ways: by the new actions *collect(p)* with precondition p and cost 0, or by the new actions *forgo(p)* with precondition $\overline{p}$, that stands for the negation of p, and cost equal to the utility $u(p)$ of p. Additional bookkeeping is needed in the translation so that these new actions can be done only after the normal actions in the original problem.

More precisely, for a STRIPS problem $P = \langle F, I, O, G \rangle$ with action costs $c(\cdot)$ and soft goals $u(\cdot)$, the equivalent, compiled STRIPS problem $P' = \langle F', I', O', G' \rangle$ with action costs $c'(\cdot)$ and no soft goals has the following components, where $F_u = \{p \mid (p \in F) \wedge (u(p) > 0)\}$ stands for the set of soft goals [Keyder and Geffner, 2009]:

- $F' = F \cup \{p' \mid p \in F_u\} \cup \{\overline{p'} \mid p \in F_u\} \cup \{normal\text{-}mode, end\text{-}mode\}$,

- $I' = I \cup \{\overline{p'} \mid p \in F_u\} \cup \{normal\text{-}mode\}$,

- $O' = O'' \cup \{collect(p), forgo(p) \mid p \in F_u\} \cup \{end\}$,

- $G' = G \cup \{p' \mid p \in F_u\}$, and

- $c'(a) = \begin{cases} c(a) & \text{if } a \in O'', \\ u(p) & \text{if } a = forgo(p), \\ 0 & \text{if } a = collect(p) \text{ or } a = end. \end{cases}$

If the STRIPS actions a are denoted as pairs $\langle Pre, Post \rangle$, where Pre stands for the preconditions of a, and $Post$ for its effects (negated atoms indicate atoms in $Del(a)$), the actions in the new compiled problem P' can be expressed as:

- $O'' = \{\langle Pre(o) \cup \{normal\text{-}mode\}, Eff(o)\rangle \mid o \in O\}$,

- $end = \langle\{normal\text{-}mode\}, \{end\text{-}mode, \neg normal\text{-}mode\}\rangle$,

- $collect(p) = \langle\{end\text{-}mode, p, \overline{p'}\}, \{p', \neg \overline{p'}\}\rangle$,

- $forgo(p) = \langle\{end\text{-}mode, \overline{p}, \overline{p'}\}, \{p', \neg \overline{p'}\}\rangle$.

The *forgo* and *collect* actions can be used only after the *end* action that makes the fluent *end-mode* true, while the actions from the original problem P can be used only when the fluent *normal-mode* is true prior to the execution of the *end* action. Moreover, exactly one of $\{collect(p), forgo(p)\}$ can appear for each soft goal p in the plan, as both delete the fluent $\overline{p'}$ which appears in their preconditions, and no action makes this fluent true. As there is no way to make *normal-mode* true again after it is deleted

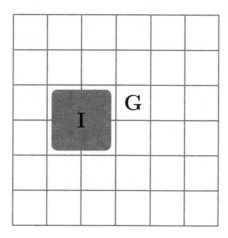

Figure 4.1: Deterministic conformant problem where a robot must move from an uncertain location I into the location with G with certainty, one cell at a time, in an $n \times n$ grid.

by the *end* action, all plans π' for P' have the form $\pi' = \langle \pi, end, \pi'' \rangle$, where π is a plan for P and π'' is a sequence of $|S'(P)|$ *collect*(p) and *forgo*(p) actions in any order, the former appearing when $\pi \models p$, and the latter otherwise.

The two problems P and P' are equivalent in the sense that there is a correspondence between the plans for P and P', and corresponding plans are ranked in the same way. More specifically, for any plan π for P, a plan π' in P' that extends π with the *end* action and a set of *collect* and *forgo* actions has cost $c(\pi') = -u(\pi) + \alpha$, where α is a constant that is independent of π and π'. Finding an optimal (maximum utility) plan π for P is therefore equivalent to finding an optimal (minimum cost) plan π' for P'. This implies that the best plans for P can be obtained from the best plans for P', and these can be computed with any optimal classical planner able to handle action costs.

From a computational point of view, the transformation above can be made more effective by means of a simple trick. Recall that for a single plan π for P, there are many extensions π' in P', all containing the same actions and having the same cost, but differing in the way the *collect* and *forgo* actions are ordered. For efficiency purposes, it makes sense to enforce a fixed but arbitrary ordering $p_1, \ldots, p_m$ on the soft goals in P by adding the dummy hard goal p_i' as a precondition of the actions *collect*(p_{i+1}) and *forgo*(p_{i+1}) for $i = 1, \ldots, m - 1$. The result is that there is a single possible extension π' of every plan π in P, and the space of plans to search is reduced. Interestingly, the cost-optimal planners that entered the Optimal Sequential Track of the 2008 IPC, fed with the translations of the problems in the Optimal Net-Benefit Track, do significantly better than the net-benefit planners that entered that track [Keyder and Geffner, 2009].

4.2 INCOMPLETE INFORMATION

Figure 4.1 shows a simple problem where an agent, whose initial state is uncertain and corresponds to one of the four shaded cells, must reach the cell marked G with certainty. For this, the agent can move one cell at a time in each one of the four directions, but cannot get passed the walls, thus any move that would take him out of the grid has no effect. The problem is very much like a classical planning problem except for the uncertain information about the initial situation. It is assumed that the problem has *uncertainty* but *no feedback* of any type; i.e., the agent does not get to observe the cell where it is located nor the nearby walls. The solution to the problem, if there is a solution, can't use feedback and hence it must be a fixed action sequence like in classical planning. The difference is that this action sequence must be applicable and achieve the goal *for any of the possible initial states*. Problems of this type are called *conformant problems*, as the solutions must conform with each possible initial state. This is a *deterministic* conformant problem as the actions have all deterministic effects. In *non-deterministic* conformant problems, solutions are also action sequences but they must conform not only with each possible initial state but with each possible non-deterministic state transition as well [Goldman and Boddy, 1996, Smith and Weld, 1998]. While conformant planning does not appear to be too interesting in itself, it is a special case of the more general problem of planning with sensing, and as we will see, the ideas developed for conformant planning provide indeed the basis for the state-of-the-art planners that sense.

The best solution to the problem in the figure is to move left two times, then to move up three times, after which the robot will know with *certainty* that it is located on the upper left-hand corner of the grid (this is sometimes called *localization*). From there, the robot can head to the goal G directly as in classical planning as all the uncertainty has been removed. We address next two issues: how to model such problems in general, and how to solve them. We then turn to a solution method for *deterministic* conformant problems that translates such problems into *classical problems*. Checking the existence of a valid plan in the conformant setting is EXPSPACE-hard, so in the worst case the translation is exponential in the number of problem variables [Bonet, 2010, Haslum and Jonsson, 1999, Rintanen, 2004a]. Still many problems exhibit a structure that makes the transformation polynomial and practical [Palacios and Geffner, 2009].

A *deterministic conformant problem* is a tuple $P = \langle F, I, O, G \rangle$ where F stands for the fluents or atoms in the problem, O stands for a set of deterministic actions a, I is a set of *clauses* over F defining the initial situation, and G is a set of literals over F defining the (conjunctive) goal. The difference to classical problems is the uncertainty in the initial situation which is described by means of clauses. Recall that a literal is an atom in F or its negation, and that a (non-empty) clause over F is a disjunction of one or more literals. We will assume that the problem is not purely STRIPS but can feature conditional effects and negation; i.e., every action a is assumed to have a precondition given by a set of fluent literals, and a set of *conditional effects* $a : C \rightarrow C'$ where C and C' are sets (conjunctions) of literals, meaning that the literals in C' become true after the action a if the literals in C were true when the action was done. The states associated with the problem P are valuations over the atoms in F, and the set of *possible initial states* are the states that satisfy the clauses in I.

The problem in the figure can be encoded by a tuple $P = \langle F, I, O, G \rangle$ where the atoms x_i and y_i encode the X and Y position of the robot in the grid, $i = 1, \ldots, n$, the goal is given by the literals x_4 and y_4, and the initial situation has clauses expressing that the different x_i and y_i atoms are mutually exclusive and that both $x_2 \vee x_3$ and $y_3 \vee y_4$ are true. The actions are four, and for example, the action *move-right* is characterized by the conditional effects:

$$move\text{-}right: \quad x_1 \longrightarrow x_2, \neg x_1 \; ;$$
$$move\text{-}right: \quad x_2 \longrightarrow x_3, \neg x_2 \; ;$$
$$\cdots$$
$$move\text{-}right: \quad x_5 \longrightarrow x_6, \neg x_5 \; .$$

A (deterministic) conformant problem $P = \langle F, I, O, G \rangle$ defines a (deterministic) conformant state model $\mathcal{S}(P)$ which is like the state model for a classical problem featuring negation and conditional effects but with *one difference*: there is no single initial state s_0 but a *set* of possible initial states S_0. A solution for P, namely a *conformant plan* for P, is an action sequence that simultaneously solves *all* the *classical state models* $\mathcal{S}'(P)$ that result from replacing the set of possible initial states S_0 in $\mathcal{S}(P)$ by each one of the states s_0 in S_0.

From a computational point of view, conformant planning can also be formulated as a *path-finding problem* over a graph, but the nodes in the graph do not represent the *states* of the problem as in classical planning, but *belief states*, where a belief state or belief is a set of states deemed possible at one point [Bonet and Geffner, 2000]. Thus, the root node of the graph is the belief $b_0 = S_0$ corresponding to the set of possible initial states, and the goal beliefs b_G are the possible non-empty sets of goal states. Likewise, the edges correspond to the belief state transitions (b, b_a) that are possible, where b_a is the belief state that results from applying the action a in the belief state b characterized as:

$$b_a = \{s' \mid s \in b \text{ and } s' \in F(a, s)\} \tag{4.2}$$

where $F(a, s)$ denotes the set of states that are possible following the action a in s. Recent proposals have advanced new heuristics for guiding the search in belief space and more compact belief state representations [Brafman and Shani, 2012b, Bryce et al., 2006, Cimatti et al., 2004, Hoffmann and Brafman, 2006, Rintanen, 2004b, To et al., 2011].

A different approach to deterministic conformant planning is based on the translation of conformant problems into classical ones [Palacios and Geffner, 2009]. The basic sound but incomplete translation removes the uncertainty in the problem by replacing each literal L in the conformant problem P by two literals KL and $K\neg L$, to be read as "L is known to be true" and "L is known to be false," respectively. If L is known to be true or known to be false in the initial situation, then the translation will contain respectively KL or $K\neg L$. On the other hand, if L is not known, then both KL and $K\neg L$ will be initially false. The result is that there is no uncertainty in the initial situation of the translation which thus represents a classical planning problem.

More precisely, the basic translation K_0 is such that if $P = \langle F, I, O, G \rangle$ is a deterministic conformant problem, the translation $K_0(P)$ is the classical planning problem $K_0(P) = \langle F', I', O', G' \rangle$ where

- $F' = \{KL, K\neg L \mid L \in F\}$

- $I' = \{KL \mid L \text{ is a unit clause in } I\}$

- $G' = \{KL \mid L \in G\}$

- $O' = O$, but with each precondition L for $a \in O$ replaced by KL, and each conditional effect $a : C \to L$ replaced by $a : KC \to KL$ and $a : \neg K\neg C \to \neg K\neg L$.[1]

The expressions KC and $\neg K\neg C$ for $C = \{L_1, L_2, \ldots\}$ are abbreviations for the conjunctions $\{KL_1, KL_2, \ldots\}$ and $\{\neg K\neg L_1, \neg K\neg L_2, \ldots\}$ respectively. Recall that in a classical planning problem, atoms that are not part of the initial situation are assumed to be initially false, so if KL is not part of I', KL will be initially false in $K_0(P)$.

---

[1] A conditional effect $a : C \to C'$ is equivalent to a collection of conditional effects $a : C \to L$, one for each literal L in C'.

The only subtlety in this translation is that each conditional effect $a : C \to L$ in P is mapped into *two* conditional effects in $K_0(P)$: a *support* effect $a : KC \to KL$, that ensures that L is known to be true when the condition C is known to be true, and a *cancellation* effect $a : \neg K \neg C \to \neg K \neg L$, that ensures that L is possible when the condition C is possible.

The translation $K_0(P)$ is *sound* as every classical plan that solves $K_0(P)$ is a conformant plan for P, but is *incomplete*, as not all conformant plans for P are classical plans for $K_0(P)$. The meaning of the KL literals follows a similar pattern: if a plan achieves KL in $K_0(P)$, then the same plan achieves L with certainty in P, yet a plan may achieve L with certainty in P without making the literal KL true in $K_0(P)$.

For completeness, the basic translation K_0 is extended into a general translation scheme $K_{T,M}$ where T and M are two parameters: a set of *tags* t and a set of *merges* m. A tag $t \in T$ is a set (conjunction) of literals L from P whose truth value in the initial situation is not known. The tags t are used to introduce a new class of literals KL/t in the classical problem $K_{T,M}(P)$ that represent the conditional statements: "if t is initially true, then L is true." Likewise, a merge m is a non-empty collection of tags t in T that stands for the Disjunctive Normal Form (DNF) formula $\bigvee_{t \in m} t$. A merge m is *valid* when one of the tags $t \in m$ must be true in I, i.e., when

$$I \models \bigvee_{t \in m} t. \tag{4.3}$$

A merge m for a literal L in P translates into a "merge action" with effects that capture a simple form of reasoning by cases:

$$\bigwedge_{t \in m} KL/t \longrightarrow KL. \tag{4.4}$$

We assume that the collection of tags T always includes a tag that stands for the empty collection of literals, called the *empty tag* and denoted as $\emptyset$. If t is the empty tag, literals KL/t are denoted as KL. The parametric translation scheme $K_{T,M}$ is the basic translation K_0 "conditioned" with the tags in T and extended with the actions that capture the merges in M. If $P = \langle F, I, O, G \rangle$ is a deterministic conformant problem, then $K_{T,M}(P)$ is the *classical planning problem* $K_{T,M}(P) = \langle F', I', O', G' \rangle$ where

- $F' = \{KL/t, K\neg L/t \mid L \in F \text{ and } t \in T\}$,

- $I' = \{KL/t \mid I, t \models L\}$,

- $G' = \{KL \mid L \in G\}$,

- $O' = \{a : KC/t \to KL/t, \ a : \neg K \neg C/t \to \neg K \neg L/t \mid a : C \to L \text{ in } P\} \cup \{a_{m,L} : [\bigwedge_{t \in m} KL/t] \to KL \mid L \in P, m \in M\}$.

As before, the literal KL is a precondition of action a in the translation $K_{T,M}(P)$ if L is a precondition of a in P. The translation $K_{T,M}(P)$ reduces to the basic translation $K_0(P)$ when M is empty (no merges) and T contains the empty tag only. Two basic properties of the general translation scheme $K_{T,M}(P)$ are that it is always *sound* (provided that merges are valid), and for suitable choice of the sets of tags and merges T and M, it is *complete*. In particular, a complete instance of the general translation $K_{T,M}(P)$ results when the set of tags T is set to the set S_0 of possible initial states of P, and

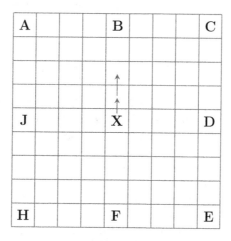

Figure 4.2: Plan Recognition: Which destination is the agent moving to after observing that he moved twice up?

a merge m is included in M such that $m = S_0$. While the resulting translation $K_{S0}(P)$ is exponential in the number of unknown atoms in the initial situation in the worst case, there is an alternative choice of tags and merges, called the $K_i(P)$ translation, that is exponential in the non-negative integer i, and that is *complete* for problems P that have a structural parameter $w(P)$, called the *width* of P, bounded by i. In problems defined over multivalued variables, this width often stands for the maximum number of variables all of which are relevant to a variable appearing in an action precondition or goal. It turns out that many conformant problems have a bounded and small width, and hence such problems can be efficiently solved by a classical planner after a low polynomial translation [Palacios and Geffner, 2009]. The conformant plans are then obtained from the classical plans by removing the "merge" actions. The translation-based approach, introduced initially for deterministic conformant planning, has been extended to deterministic planning with sensing [Albore et al., 2009, Bonet and Geffner, 2011, Brafman and Shani, 2012a,b]. In Chapter 5, we will look at a related notion of width in the more general setting of non-deterministic planning.

4.3   PLAN AND GOAL RECOGNITION

The need to recognize the goals and plans of an agent from observations of his behavior arises in a number of tasks. Plan recognition is like planning but in reverse: while in planning the goal is given and a plan is sought, in plan recognition, part of a plan is observed, and the agent goal is sought [Geib and Goldman, 2009, Kautz and Allen, 1986, Yang, 2009]. Figure 4.2 shows a simple scenario of plan recognition where an agent is observed to move up twice from cell X. The question is which is the most likely destination among the possible targets A to J. Clearly, A, B, and C appear to be more likely destinations than D, E, or F. The reason is that the agent is moving away from these other targets, while it's not moving away from A, B, or C. The second question is whether B can be regarded as more likely than A or C. There are indeed good reasons for this. If we adopt a Bayesian formulation, the probability of a hypothesis H given the observation Obs, $P(H|Obs)$ is given by the formula [Pearl, 1988]:

$$P(H|Obs) = \frac{P(Obs|H)\,P(H)}{P(Obs)} \qquad (4.5)$$

where $P(Obs|H)$ represents how well the hypothesis H predicts the observation Obs, $P(H)$ stands for how likely is the hypothesis H a priori, and $P(Obs)$, which affects all hypotheses H equally, measures how surprising is the observation. In our problem, the hypotheses are about the possible destinations of the agent, and since there are no reasons to assume that one is more likely a priori than the others, Bayes' rule yields that $P(H|Obs)$ should be proportional to the likelihood $P(Obs|H)$ that measures *how well H predicts Obs*. Going back to the figure, and assuming that the agent is reasonably "rational" and hence wants to achieve his goals with least cost, it's clear that A, B, and C predict Obs better than D, E, F, and also that B predicts Obs better than A and C. This is because there is a single optimal plan for B that is compatible with Obs, but there are many optimal plans for A and for C, some of which are not compatible with Obs (as when the agent moves first left or right, rather than up). We say that a plan π is compatible with the observed action sequence Obs when the action sequence Obs is embedded in the action sequence π, i.e., when Obs is π but with certain actions in π omitted (not observed).

The reasoning above reduces goal recognition to Bayes' rule and how well each of the possible goals predicts the observed action sequence. Moreover, how well a goal G predicts the sequence Obs turns out to depend on considerations having to do with *costs*, and in particular, *two cost measures:* the cost of achieving G through a plan *compatible* with the observed action sequence Obs, and the cost of achieving G through a plan that is *not* compatible with Obs. We will denote the first cost as $c_P(G + Obs)$ and the second cost as $c_P(G + \overline{Obs})$, where P along with the observations Obs define the *plan recognition problem*. That is, P is like a classical planning problem but with the actual goal hidden and replaced by a set $\mathcal{G}$ of possible goals G, i.e., $P = \langle F, I, O, \mathcal{G} \rangle$. The plan recognition problem is to infer the probability distribution $P(G|Obs)$ over the possible goals $G \in \mathcal{G}$, where each possible goal G can be a (conjunctive) set of atoms.

For the plan recognition problem in Figure 4.2, the measures $c_P(B + Obs)$ and $c_P(B + \overline{Obs})$, encoding the costs of getting to B from X through plans compatible and incompatible with the observed action sequence Obs, are 4 and 6 respectively, assuming moves in each one of the four possible directions, each with cost 1. On the other hand, the pairs of measures $(c_P(G + Obs), c_P(G + \overline{Obs}))$ for G equal to A, J, and H, are $(8, 8)$, $(8, 4)$, $(12, 8)$ respectively.

The key feature is actually the *cost difference* $\Delta(G, Obs) = c_P(G + \overline{Obs}) - c_P(G + Obs)$ for each goal G which can range from $-\infty$ to $+\infty$. It can be argued that the higher the value of $\Delta(G, Obs)$, the better that G predicts Obs, and hence the higher the likelihood $P(Obs|G)$. In particular, $\Delta(G, Obs)$ is ∞ when *all* the plans for G comply with Obs, and $-\infty$ when *none* of them complies with Obs. Values in the middle reflect how good are the plans that comply and do not comply with the observed action sequence Obs. In our example, $\Delta(G, Obs)$ is 2 for $G = $ B, 0 for $G = $ A and $G = $ C, and -4 for the other possible goals. Hence $P(Obs|G)$ is largest for $G = $ B, smaller for $G = $ A and $G = $ C, and smallest for the rest. The function used by Ramírez and Geffner [2010] for mapping the cost difference $\Delta(G, Obs) = c_P(G + \overline{Obs}) - c_P(G + Obs)$ into the likelihoods $P(Obs|G)$ is the *sigmoid function*:

$$P(Obs|G) = \frac{1}{1 + e^{-\beta \Delta(G, Obs)}} \qquad (4.6)$$

where β is a positive constant. This expression is derived from the assumption that while the observed agent is not perfectly rational, he is more likely to follow cheaper plans, according to a Boltzmann distribution. The larger the value of the constant β, the more rational the agent, and the less likely that he will follow suboptimal plans.

The target distribution $P(G|Obs)$ over the possible goals $G \in \mathcal{G}$ given the observation sequence Obs can thus be obtained in three steps. First, the costs $c_P(G + Obs)$ and $c_P(G + \overline{Obs})$ of achieving each possible goal G with plans that are compatible and incompatible with the observed action sequence Obs are determined. Then, the resulting cost differences $\Delta(G, Obs)$ are plugged into Eq. 4.6 to yield the likelihoods $P(Obs|G)$. Finally, these likelihoods are plugged into Bayes' rule (4.5) from which the goal posterior probabilities are obtained. The probabilities $P(Obs)$ used in Bayes' rule are obtained by normalization (goal probabilities must add up to 1 when summed over all possible goals).

The open question is how to compute the cost measures $c_P(G + Obs)$ and $c_P(G + \overline{Obs})$. Ramírez and Geffner [2010] show that these costs correspond to the costs of two *classical planning problems*, that we will call $P(G + Obs)$ and $P(G + \overline{Obs})$, defined from the plan recognition problem $P = \langle F, I, O, \mathcal{G} \rangle$, where $\mathcal{G}$ stands for the set of possible agent goals, and the observed action sequence Obs. If we assume that no action occurs twice in the observed sequence Obs, the problems $P(G + Obs)$ and $P(G + \overline{Obs})$ are like P but with extra atoms p_a for each $a \in Obs$, all initially false, such that p_a is made into an effect of the action a when a is the first action in Obs, while $p_b \rightarrow p_a$ is made into a conditional effect of a, when b is the action that immediately precedes a in sequence Obs. The cost $c_P(G + Obs)$ is then the cost of this classical problem for the goal $G' = G \cup \{p_a\}$, where a is the last action in the sequence Obs, and the cost $c_P(G + \overline{Obs})$ is the cost of the same classical problem but with goal $G'' = G \cup \{\neg p_a\}$ where $\neg p_a$ is the negation of p_a. In other words, the constraint of achieving a possible goal G in a way that is compatible or incompatible with an observed action sequence Obs, is mapped into the problem of achieving G and a suitable dummy goal associated with Obs in a transformed classical problem.

Figure 4.3 shows a slightly different example, where the path followed by the agent is shown on the left as time progresses. The curves on the right show the resulting goal posterior probabilities over each one of the possible targets as a function of time. The account presented is not tied to agents navigating in grids but is completely domain-independent. For computing the posterior probabilities, $2 \times |\mathcal{G}|$ classical planning problems need to be solved. These probabilities will be exact if the problems are solved optimally, and will be approximate if they are solved with more scalable non-optimal planners. The use of model-based approaches to *behavior generation* for the inverse task of *behavior recognition* has been considered recently for other models such as MDPs [Baker et al., 2009] and POMDPs [Ramírez and Geffner, 2011]. Moreover, the approaches can also be used to recognize both goals and agent beliefs, by just replacing the set of possible goals by a set of possible goals and initial belief *pairs*.

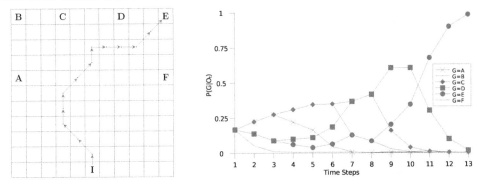

Figure 4.3: *Left:* Red path shows noisy walk of agent Obs_t as time t progresses. *Right:* Curves show goal posterior probabilities $P(G|Obs_t)$ for each possible target as a function of time.

4.4 FINITE-STATE CONTROLLERS

Finite-state controllers represent an action selection mechanism widely used in video games and mobile robotics. In comparison to plans and POMDP policies, to be studied later, finite-state controllers have two advantages: they are often extremely compact, and they are general, applying not just to one problem but to many variations as well. As an illustration, Figure 4.4(a) depicts a simple problem over a 1×5 grid where a robot, initially at one of the two leftmost positions, must visit the rightmost position, marked B, and get back to A. Assuming that the robot can observe the mark in the current cell if any, and that the actions *Left* and *Right* deterministically move the robot one unit left and right respectively, the problem can be solved by planners that sense and POMDP planners. A solution to the problem, however, can also be expressed as the finite-state controller shown on the right. Starting in the controller state q_0, this controller selects the action *Right*, whether A or no mark ('−') is observed, until observing B. Then the controller selects the action *Left*, switches to state q_1, and remains in this state selecting the action *Left* as long as no mark is observed. Later, when a mark is observed, no further actions are taken as the agent must be back at A, having achieved the goal.

The finite-state controller displayed in the figure has two appealing features: it is very compact (it involves two states only), and it is very general. Indeed, the problem can be changed in a number of ways and the controller would still work, driving the agent to the goal. For example, the *size of the grid* can be changed from 1×5 to $1 \times n$, the agent can be placed *initially* anywhere in the grid (except at B), and the actions can be made *non-deterministic* by adding "noise" so that the agent can move one or two steps at a time. The controller would work for all these variations. This generality is well beyond the power of plans or policies that are normally tied to a particular state space. Memoryless controllers or policies [Littman, 1994] are widely used as well, and they are nothing but finite-state controllers with a single controller state. Additional states provide controllers with a memory that allows different actions to be selected given the same observation.

The benefits of finite-state controllers, however, come at a price: unlike plans, they are usually not derived automatically from a model but are written by hand—a task that is non-trivial even in the simplest cases. Recently, however, the problem of deriving compact and general finite-state controllers using planners has been considered [Bonet et al., 2009]. Once again, this is achieved by using classical planners over suitable transformations. We sketch the main ideas below.

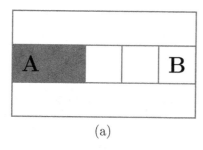 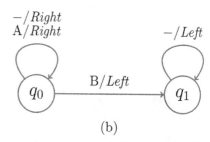

(a)   (b)

Figure 4.4: (a) A partially observable problem where an agent initially in one of the two leftmost positions has to go to the cell marked B and then back to the cell marked A. These two marks are observable. (b) A 2-state controller that solves this problem and many variations of it. The circles are the controller states, and an edge $q \rightarrow q'$ labeled o/a means to perform action a when the observation is o in state q, switching then to state q'. The initial controller state is q_0.

A finite-state controller C_N with N controller states $q_0, \ldots, q_{N-1}$, for a partially observable problem P with possible observations $o \in O$ is fully characterized by the tuples (q, o, a, q') associated with the edges $q \xrightarrow{o/a} q'$ in the controller graph. These edges, and hence these tuples, prescribe the action a to do when the controller state is q and the observation is o, switching then to the controller state q' (which may be equal to q or not). A controller solves P if starting in the distinguished controller state q_0, all the executions that are possible given the controller reach a goal state. The key question is how to find the tuples (q, o, a, q') that define such a controller. In the approach by Bonet et al. [2009], the problem P is transformed into a problem P' whose actions are associated with each one of the possible tuples (q, o, a, q'), and where extra fluents p_q and p_o for keeping track of the controller states and observations are introduced. The action $\langle t \rangle$ associated with the tuple $t = (q, o, a, q')$ behaves then very much like the action a but with two differences: first, the atoms p_q and p_o are added to the body of each conditional effect, so that the resulting action $\langle t \rangle$ behaves like the original action a but only when the controller state is q and the observation is o; second, the action makes the atom p_q false and the atom $p_{q'}$ true, in accordance with the interpretation of the tuple (unless $q = q'$). Additional bookkeeping is required in the transformed problem P' to prevent plans from executing actions $\langle t \rangle$ and $\langle t' \rangle$ when $t = (q, o, a, q')$, $t' = (q, o, a', q'')$, and $a \neq a'$ or $q' \neq q''$. The reason is that no controller can include such pairs of tuples, as the action and new controller state are always a *function* of the current controller state and observation. Interestingly, the transformation from P into P' eliminates sensing by making the effects of the actions conditional on the current controller state and observation. The result is that while P is a partially observable problem, P' is a *conformant* problem, which as we have seen in Section 4.2, can be further transformed into a classical problem. The actions $\langle t \rangle$ that solve such classical problems encode the tuples that define the controller with up to N states that solves P.

As a further illustration of the power of these transformations, Figure 4.5, on the left, shows a problem inspired in the use of deictic representations [Ballard et al., 1997, Chapman, 1989], where a visual-marker (the circle on the lower left) must be placed on top of a green block by moving it one cell at a time. The location of the green block is not known, and the observations are whether the cell currently marked contains a green block (G), a non-green block (B), or neither (C), and whether

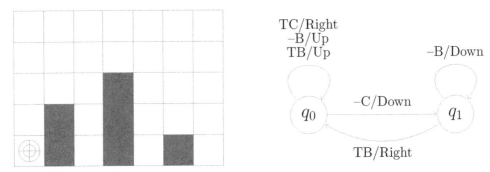

Figure 4.5: *Left:* Problem where a visual-marker (mark on the lower left cell) must be placed on top of a green block whose location is not known, by moving the mark one cell at a time, and by observing what's in the marked cell. *Right:* Finite-state controller obtained with a classical planner from suitable translation. The controller solves the problem and any variation resulting from changes in either the number or configuration of blocks.

this cell is at the level of the table (T) or not (–). The finite-state controller shown on the right has been computed by running a *classical planner* over a translation obtained following the two steps above: one, from the original partially observable problem into a conformant problem; the second, from the conformant problem into a classical one. The solution to the classical problem represents the finite-state controller that is shown on the right. Interestingly, this controller not only solves the problem shown on the left, but the reader can verify that it also solves *any modification in the problem resulting from changes in either the dimensions of the grid, number or configuration of blocks* [Bonet et al., 2009]. This is quite remarkable and illustrates that the combined use of transformations and classical planners can be very powerful indeed.

4.5   TEMPORALLY EXTENDED GOALS

Classical planning is about acting on a system to drive it into a final state where a goal holds. Such tasks are sometimes called "reachability" problems. In the last few years, *temporally extended goals* expressed in temporal logics have been increasingly used to capture a richer class of plans where restrictions over the whole sequence of states must be satisfied as well [Bertoli et al., 2003, de Giacomo and Vardi, 1999, Gerevini et al., 2009]. A temporally extended goal may state, for example, that any borrowed tool should be kept clean until it is returned, defining a constraint that does not apply to a single state but to a whole state sequence. A plan achieves a goal while satisfying a state-trajectory constraint when the plan achieves the goal in the standard sense, and in addition, the state sequence that it generates satisfies the constraint.

A standard language for expressing trajectory constraints is Linear Temporal Logic or LTL, a logic originally proposed as a specification language for concurrent programs [Pnueli, 1977]. Formulas of LTL are built from a set F of atoms and are closed under the boolean operators, the unary temporal operators $\bigcirc$, $\Diamond$, and $\Box$, and the binary temporal operator $\mathcal{U}$. Intuitively, $\bigcirc\varphi$ says that φ holds at the *next* instant, $\Diamond\varphi$ says that φ will *eventually* hold at some future instant, $\Box\varphi$ says that from the current instant on φ will *always* hold, and $\varphi\mathcal{U}\psi$ says that at some future instant ψ will hold and *until* that point φ holds. As an example, the formula $\Box(p \supset \bigcirc q)$ says that if p is true at any time point, then q must be true at the following time point.

The semantics of LTL is given in terms of *infinite* state sequences $\sigma = s_0, s_1, \ldots, s_i, \ldots$ where the indices i stand for time points, and the state s_i represents a truth valuation over F at time i. If we let $\sigma(i)$ stand for the state s_i in the sequence σ, the conditions under which a state sequence σ satisfies an arbitrary LTL formula φ at time i, written $\sigma, i \models \varphi$, can be given inductively as follows:

- $\sigma, i \models p$, for $p \in F$, iff $p \in \sigma(i)$.

- $\sigma, i \models \neg\varphi$ iff not $\sigma, i \models \varphi$.

- $\sigma, i \models \varphi \wedge \varphi'$ iff $\sigma, i \models \varphi$ and $\sigma, i \models \varphi'$.

- $\sigma, i \models \bigcirc\varphi$ iff $\sigma, i+1 \models \varphi$.

- $\sigma, i \models \Diamond\varphi$ iff for some $j \geq i$, we have that $\sigma, j \models \varphi$.

- $\sigma, i \models \Box\varphi$ iff for all $j \geq i$, we have that $\sigma, j \models \varphi$.

- $\sigma, i \models \varphi \mathcal{U} \varphi'$ iff for some $j \geq i$, we have that $\sigma, j \models \varphi'$ and for all $k, i \leq k < j$, we have that $\sigma, k \models \varphi$.

A formula φ *is true* or *satisfied* in σ, written $\sigma \models \varphi$, if $\sigma, 0 \models \varphi$. For determining whether a given plan $\pi = a_0, \ldots, a_{n-1}$ for a classical planning problem $P = \langle F, I, O, G \rangle$ satisfies a temporally extended goal expressed as an LTL formula over F, it is normally assumed that the finite state sequence $s_0, \ldots, s_n$ generated by the plan π represents the *infinite state sequence $s_0, \ldots, s_n, s_n, s_n, \ldots$* where the last state s_n in the sequence is repeated forever [Bacchus and Kabanza, 2000]. This is an assumption that can be used for many LTL formulas but not for all, as some formulas may be satisfiable but not by sequences of this type. A formula like $\Box(\Diamond at(p1) \wedge \Diamond at(p2))$ expressing that from any time point on, the robot has to be eventually at position 1 and eventually at position 2, is one such example. These formulas require the consideration of more general infinite state sequences where one finite sequence $s_0, \ldots, s_n$ is followed by another finite state sequence $s_n, s'_1, \ldots, s'_m, s_n$ that forms a loop and is repeated infinitely often, and where the states s'_i are different than s_n. We'll focus now on the fragment of temporally extended goals expressed in LTL where "completed" state sequences $s_0, \ldots, s_n, s_n, s_n, \ldots$ suffice. Following Bauer and Haslum [2010], we refer to this as the *infinite-extension* semantics for LTL, or simply the *IE-semantics*.

We turn thus to the problem of *computing* a finite plan $\pi = a_0, \ldots, a_{n-1}$ for a classical planning problem $P = \langle F, I, O, G \rangle$ such that the completed infinite state sequence $s_0, \ldots, s_n, s_n, s_n, \ldots$ that results from the plan satisfies an LTL formula φ. It turns out that this problem can be solved by mapping the classical problem P and the formula φ into a new classical planning problem P_φ whose solutions represent plans for P that satisfy φ [Baier et al., 2009, Cresswell and Coddington, 2004, Edelkamp, 2006]. Rather than focusing on the syntactic details of the translation, we describe the main idea semantically.

We know by now that the planning problem $P = \langle F, I, O, G \rangle$ represents a state model $\mathcal{S}(P) = (S, s_0, S_G, A, f)$, that can also be understood as a deterministic finite automaton $A^P = (\Sigma^P, Q^P, q_0^P, \delta^P, F^P)$ where the input alphabet is $\Sigma^P = O$, the states are $Q^P = S$, the initial state is $q_0^P = s_0$, the transition function δ^P is such that $s' \in \delta^P(a, s)$ iff $s' = f(a, s)$, and the accepting states are $F^P = S_G$. The LTL formula φ defines in turn a non-deterministic Büchi automaton

$A^\varphi = (\Sigma^\varphi, Q^\varphi, q_0^\varphi, \delta^\varphi, F^\varphi)$ where the input alphabet is $\Sigma^\varphi = S$, and the accepted inputs are the infinite state sequences that satisfy φ, defined as the inputs that generate state sequences over Q^φ that pass through accepting states in F^φ infinitely often [Gerth et al., 1995, Vardi and Wolper, 1994]. Under the IE-semantics, however, it is enough to reach an accepting state once, and hence the automaton A^φ can be regarded as a standard non-deterministic finite automaton, which can be *determinized* using standard methods [Hopcroft and Ullman, 1979, Sipser, 2006].

Therefore, under the IE-semantics, the valid plans that satisfy an LTL formula φ are the action sequences $\pi = a_0, \ldots, a_{n-1}$ that generate state sequences $\sigma = s_0, \ldots, s_n$ such that π is accepted by the first automaton A^P and σ is accepted by the second automaton A^φ. Thus, the classical planning problem P_φ whose solutions encode the plans for P that satisfy the LTL formula φ can be expressed as the compact representation of the *product* of two deterministic automata: the deterministic automaton A^P associated with the problem P, and the deterministic version of the automaton A^φ associated with the LTL goal φ. The states over the problem P_φ, which represent the truth-valuations over the atoms in P_φ, stand for pairs (s, q) where s captures the state on the first automaton and q captures the state on the second automaton. This construction requires the addition of atoms p_q in P_φ for such states q, in addition to the atoms in P. The actions in P_φ are the actions in P but with effects on the atoms p_q in correspondence with the second automaton. Likewise, the initial state of P_φ is the initial state of P extended with the atom p_{q_0}, and the goal in P_φ is the goal of P conjoined with the disjunction of atoms p_q for accepting states q. Approaches and transformations for dealing with *arbitrary LTL goals*, that may require *plans with loops*, and "lasso" state sequences have also been developed [Albarghouthi et al., 2009, Kabanza and Thiébaux, 2005, Patrizi et al., 2011, 2013].

CHAPTER 5

Planning with Sensing: Logical Models

In this chapter we focus on models and methods for planning with uncertainty and sensing. This is usually called *partial observable planning*, *planning with sensing*, or *contingent planning*. In these models the true state of the environment is not assumed to be known or predictable, yet partial information about the state is assumed to be available from sensors. Uncertainty is represented by *sets* of states, referred to as *beliefs*. We will then consider probabilistic models where beliefs are not represented by *sets* of states but by *probability distributions*. Logical and probabilistic models however are closely related. A key difference is that, in the absence of probabilistic information, policies or plans are evaluated by their *cost in the worst case* rather than their *expected cost*. There may indeed be policies with small expected cost to the goal but infinite cost in the worst case, as when the state trajectories that fail to reach the goal in a bounded number of steps have a vanishing small probability. Still, as we will see, the policies that ensure that the goal is achieved with certainty can be fully characterized in the logical setting without probabilities at all as the policies that are strongly cyclic [Daniele et al., 1999].

In this chapter we consider a general model for planning with sensing, a language for expressing these models in compact form, the notion of a solution, policy, or plan for such models, and offline and online algorithms for action selection. Since all these algorithms require keeping track of beliefs, we then present methods for tracking beliefs that exploit the structure of the problem and are exponential in a problem width parameter. We also review strong cyclic policies and methods for computing them.

5.1 MODEL AND LANGUAGE

The model for planning with sensing extends the model for (non-deterministic) conformant planning with a sensor model. More precisely, the state model for *conformant planning* is a tuple of the form $S = \langle S, S_0, S_G, A, F \rangle$ where

- S is a finite state space,

- S_0 is a non-empty set of possible initial states, $S_0 \subseteq S$,

- S_G is a non-empty set of goal states, $S_G \subseteq S$,

- A is a set of actions, with $A(s)$ denoting the sets of actions applicable at $s \in S$,

- F is a non-deterministic state-transition function such that $F(a, s)$ denotes the non-empty set of possible successor states that follow action a in s, for $a \in A(s)$, and

- $c(a, s)$ are positive action costs for $s \in S$ and $a \in A(s)$.

The model extends the classical planning model by allowing uncertainty in the initial situation and in the transition function. A solution to a conformant model is an action sequence that maps each possible initial state into a goal state. More precisely, $\pi = \langle a_0, \ldots, a_{n-1} \rangle$ is a *conformant plan* if for each possible sequence of states $s_0, s_1, \ldots, s_n$ such that $s_0 \in S_0$ and $s_{i+1} \in F(a_i, s_i)$, $i = 0, \ldots, n-1$, action a_i is applicable in s_i and s_n is a goal state.

Conformant planning can be cast as a path-finding problem over a graph whose nodes are *beliefs states*: sets of states that the agent deems possible at one point. The initial node is the initial belief state $b_0 = S_0$ and the target nodes are the goal beliefs b_G, non-empty sets of goal states $s \in S_G$. The actions a, whether deterministic or not, map a belief state b into the belief state b_a:

$$b_a = \{s' \mid \text{there is a state } s \text{ in } b \text{ such that } s' \in F(a, s)\}. \tag{5.1}$$

For the resulting paths to encode conformant plans, an action a must be regarded as applicable in the belief state b, written $a \in A(b)$, when a is applicable in each state s in b, or equivalently, when the preconditions of a are true (in all the states) in b.

The model for *planning with sensing* $S = \langle S, S_0, S_G, A, F, O \rangle$ is the model for conformant planning extended with a *sensing model* O: a function $O(s, a)$ that maps state-action pairs into non-empty sets of observation tokens. The expression $o \in O(s, a)$ means that o is a possible observation token when s is the true state of the system and a is the last action done. That is, every time that the agent executes the action a resulting in the state s, the agent gets an observation token from $O(s, a)$. This observation o provides partial information about the true but possibly hidden state s, since it rules out states for which the observation token o is not possible, i.e., the states s' for which $o \notin O(s', a)$. If two different observations belong to $O(s, a)$, then either one can be observed in s when a is the last action. We say that the sensing is *deterministic* or *noiseless* when $O(s, a)$ is a singleton for every pair (s, a), else it is *non-deterministic* or *noisy*.

If the belief state for the agent is b and the observation o is obtained after applying the action a in b, the new belief state, denoted as b_a^o, is given by the states in b_a that are compatible with o:

$$b_a^o = \{s \mid s \in b_a \text{ and } o \in O(s, a)\}. \tag{5.2}$$

An observation o is *possible* in a belief state b_a if $o \in O(s, a)$ for some state s in b_a. Alternatively, the observation o is possible in b_a if and only if the resulting belief state b_a^o is not empty.

LANGUAGE

Conformant models can be expressed in compact form through a set of *state variables*. For convenience, in the partially observable setting, we assume that these variables are not necessarily boolean. More precisely, a *conformant planning problem* is a tuple $P = \langle V, I, A, G \rangle$ where V stands for the problem variables X, each one with a finite and discrete domain D_X, I is a set of clauses over the V-literals defining the initial situation, A is set of actions, and G is a set of V-literals defining the goal. Every action a has a precondition $Pre(a)$ given by a set of V-literals, and a set of conditional effects $a : C \rightarrow E_1 | \ldots | E_n$, where C and each E_i is a set (conjunction) of V-literals. The conditional effect is *non-deterministic* if $n > 1$. A non-deterministic action is an action with one or more non-deterministic effects.

The *conformant problem* $P = \langle V, I, A, G \rangle$ defines the conformant model $\mathcal{S}(P) = \langle S, S_0, S_G, A, F \rangle$, where S is the set of possible valuations over the variables in V, S_0 and S_G are the set of valuations that satisfy I and G respectively, $A(s)$ is the set of actions whose preconditions are true in s, and $F(a, s)$ is a non-deterministic state transition function where $s' \in F(a, s)$ is a possible successor state of action a in state s for $a \in A(s)$. The set $F(a, s)$ of such possible successors s' is defined by the conditional effects $a : C \rightarrow E_1 | \ldots | E_n, n \geq 1$, whose body C is true in s. Basically, any logically consistent choice σ of heads E_i, one for each conditional effect whose body is true in s must define a *deterministic* transition function $f_\sigma(a, s)$. $F(a, s)$ is the *non-deterministic* transition function that results from collecting all the successor states s' that are possible given any of these deterministic functions.

A *partially observable problem* P is a tuple $P = \langle V, I, A, G, V', W \rangle$ that extends the description $\langle V, I, A, G \rangle$ of a conformant model with a compact encoding of a *sensor model*. This sensor model is defined syntactically by means of a set V' of *variables* Y with a finite domain D_Y that are assumed to be *observable*, and a set W of *formulas* $W_a(Y = y)$ over the state variables V of the problem that determine the states s over which the atom $Y = y$ may be observed. More precisely, the sensor model $O(s, a)$ defined by W is such that $o \in O(s, a)$ iff o is a valuation over the observable variables $Y \in V'$ such that $Y = y$ is true in o only if the formula $W_a(Y = y)$ is true in s for $y \in D_Y$. In other words, an observation o represents a maximal consistent set of partial observations $Y = y$ where Y is an observable variable and y is a possible value of Y. Such an observation o is possible in the state s after doing action a if the formulas $W_a(Y = y)$ are all true in s.

Two last remarks. First, some of the state variables X may be observable and hence belong to both V and V'. In such a case, the formula $W_a(X = x)$ for the different actions and possible values of X is given by $X = x$. Second, the formulas $W_a(Y = y)$ for the different values y in D_Y must be logically *exhaustive*, as every state-action pair must give rise to some observation over each observable variable Y. If in addition, the formulas $W_a(Y = y)$ for the different values y are logically *exclusive*, every state-action pair gives rise to a single observation $Y = y$ and the sensing over Y is deterministic.

As an example, if X encodes the location of an agent, and Y encodes the location of an object that can be seen by the agent when $X = Y$, we can have an observable variable $Z \in \{Yes, No\}$ encoding whether the object is seen by the agent or not, with observation model $W_a(Z = Yes)$ given by $\bigvee_{l \in D}(X = l \wedge Y = l)$, where D is the set of possible locations and a is any action, and $W_a(Z = No)$ given by the negation of this formula. The resulting sensor is deterministic. A non-deterministic sensor could be used if, for example, the agent cannot detect with certainty the presence of the object at some other locations $l \in D'$. For this, $W_a(Z = Yes)$ and $W_a(Z = No)$ can be set to the disjunction of their previous expression and the formula $\bigvee_{l \in D'}(X = l)$. The result is that the two observations $Z = Yes$ and $Z = No$ will be possible in the states where the agent is at some location $l \in D'$, whether the object is in the same location or not.

5.2   SOLUTIONS AND SOLUTION FORMS

An *execution* for a partially observable problem is an interleaved sequence of actions a_k and observations o_k, $\langle a_1, o_1, a_2, o_2, \ldots \rangle$. An execution may be finite or infinite. A finite execution $\langle a_1, o_1, \ldots, a_i, o_i \rangle$ is also called a *history*. Associated with a finite execution or history $h_i = \langle a_1, o_1, \ldots, a_i, o_i \rangle$, there is a belief b_i. For the empty history h_0, the belief is the initial belief b_0, while for the history $h_{i+1} = h_i, a_{i+1}, o_{i+1}$, the resulting belief b_{i+1} is obtained from the belief b_i associated with the history h_i, the action a_{i+1}, and the observation o_{i+1}, using Eq. 5.2 for the belief b_a^o, with $b = b_i$, $a = a_{i+1}$, and $o = o_{i+1}$.

The executions that are *possible* are the ones where the actions are applicable and the observations are possible. More precisely, the possible executions or histories $\langle a_1, o_1, a_2, o_2, \ldots \rangle$ are defined recursively: the empty history h_0 is possible, and if history h_i is possible, history $h_{i+1} = h_i, a_{i+1}, o_{i+1}$ is possible iff the action a_{i+1} is applicable in the belief b_i that results from the history h_i, and the observation o_{i+1} is possible in the belief b_a for $b = b_i$ and $a = a_{i+1}$.

The *solution* of a planning problem with sensing is a choice of actions that ensures that all the resulting executions reach a goal belief in a finite number of steps. We make this precise below. We consider two different types of *solution forms*. In both cases, the choice of the action to do depends on the past actions and observations. In the first case, the choice is expressed as a function mapping *histories* into actions; in the second, as a function mapping *belief states* into actions. We will refer to these functions as *control policies* or simply as *policies*, and denote them with the symbol π. Partially observable problems can also be solved by means of *finite-state controllers*, but we will not delve further on such an alternative solution form, which can be more compact but is harder to derive (yet see Section 4.4).

One difference between history-based and belief-based policies is that the set of histories cannot be bounded a priori, while the set of beliefs is large but bounded: exponential in the number of states. On the other hand, actions define a graph over histories that is acyclic (histories can only grow in size), while the graph over beliefs can be cyclic. These two features imply that different algorithms may be convenient for computing one type of policy or the other.

We focus on policies π where $\pi(h_i)$ and $\pi(b_i)$ express actions a that are applicable in the history h_i or belief b_i respectively. An action is applicable in a history h if it is applicable in the belief that results from this history. The functions π, however, can be *partial* and don't have to be defined for all possible inputs. We write $\pi(h) = \perp$ and $\pi(b) = \perp$ to express that policy π is *undefined* for the history h or belief b respectively.

The executions that are possible *given a policy* π, whether history or belief-based, are defined as follows: the empty execution h_0 is always possible given π, and if h_i is a possible execution given π, the execution h_i, a_{i+1}, o_{i+1} is possible given π iff 1) a_{i+1} is the action dictated by the policy π in h_i, and 2) o_{i+1} is a possible observation after action a_{i+1} is applied in the belief b_i that results from h_i. A possible execution h given a policy π is *complete* either if it is *infinite* or if $\pi(h) = \perp$ or $\pi(b) = \perp$, where b is the belief that results from the history h.

Finally, a policy π *solves* a partially observable problem P iff all the complete executions that are possible given π are *finite*, and terminate in a *goal belief*. Assuming that the cost of actions is uniform and equal to 1, the *cost* of a policy π is defined as the *number of actions* in the longest execution that is possible given π. The *optimal policies* that solve P are the ones that minimize the cost of achieving the goal in the worst case.

A belief-based policy π that solves a problem P induces a unique history-based policy π' that solves P where $\pi'(h) = \pi(b)$ when b is the belief that results from history h. On the other hand, a history-based policy π induces a unique belief-based policy π' only when there are no two histories h and h' with the same associated belief b such that $\pi(h) \neq \pi(h')$. Still, if a problem has a solution in one form, it certainly has a solution in the other form.

EXAMPLE

Let us consider a problem where there is a toy in one of two closed boxes, the goal is to have the toy, and the actions are to open a box, to inspect an open box, and to pick up the toy from a box if the box is open and contains the toy. The problem can be modeled by boolean state variables for encoding that a box is open ($opened(box)$), the toy is in a box ($in(toy, box)$), the contents of a box are visible ($visible(box)$), and the toy is being held ($hold(toy)$). These variables are all false in the initial situation, except for the variables $in(toy, box), box \in \{box_1, box_2\}$, which are not known initially. What is known instead is that either $in(toy, box_1)$ or $in(toy, box_2)$ is true. In addition, an observable variable Y with three possible values $\{yes, no, ?\}$ can be used to model whether the toy is seen in a box. For this, we can set the observable model formula $W_a(Y = yes)$ to $in(toy, box) \wedge visible(box)$, $W_a(Y = no)$ to $\neg in(toy, box) \wedge visible(box)$, for $a = inspect(box)$ and $box \in \{box_1, box_2\}$, and both formulas to $false$ when $a \neq inspect(box)$. Likewise, $W_a(Y = ?)$ is $\neg visible(box)$ for $a = inspect(box)$, and $true$ otherwise. The action that makes the atom $visible(box)$ true is $inspect(box)$ whose precondition is that the box is open. The actions of opening a box and picking up the toy from a box have preconditions $\neg opened(box)$ and $in(toy, box) \wedge opened(box)$ respectively. A strategy for solving the problem is to open box 1, inspect its contents, pick up the toy if there, and else, open box 2 and pick up the toy from box 2. This strategy can be captured by the history-based policy π where:

$$\pi(h_0) = open(box_1) \text{ for the empty history } h_0 = \langle\rangle,$$

$$\pi(h_1) = inspect(box_1) \text{ for } h_1 = \langle open(box_1), Y = ?\rangle,$$

$$\pi(h_2) = pickup(toy, box_1) \text{ for } h_2 = \langle open(box_1), Y = ?, inspect(box_1), Y = yes\rangle,$$

$$\pi(h_2') = open(box_2) \text{ for } h_2' = \langle open(box_1), Y = ?, inspect(box_1), Y = no\rangle,$$

$$\pi(h_3') = pickup(toy, box_2) \text{ for } h_3' = \langle h_2', open(box_2), Y = ?\rangle.$$

No other executions are possible given this policy. Initially the belief state contains two states differing only in the truth of the atoms $in(toy, box_1)$ and $in(toy, box_2)$ After the action $inspect(box_1)$ and the resulting observation $Y = yes$ or $Y = no$, the belief state reduces to a single state where $in(toy, box_1)$ and $in(toy, box_2)$ are true respectively. This is because for $a = inspect(box_1)$, the sensor model $O(s, a)$ is such that $o = (Y = yes)$ can be observed only when the state s satisfies the formula $W_a(Y = yes) = in(toy, box_1) \wedge visible(box_1)$, while $o = (Y = no)$ can be observed only when s satisfies the formula $W_a(Y = no) = \neg in(toy, box_1) \wedge visible(box)$. Thus, after the history h_2, $in(toy, box_1)$ must be true, while after the history h_2', $in(toy, box_1)$ must be false. Since $in(toy, box_1)$ is false only in the state where $in(toy, box_2)$ is true, it follows that $in(toy, box_2)$ must be true after h_2'.

5.3   OFFLINE SOLUTION METHODS

As in classical planning, we consider two types of computational methods for partial observable planning: *offline* methods that produce policies that solve the problem, and *online* methods that select the action to do next without solving the whole problem first. Clearly, online methods are more practical and scale up better than offline methods but do not have the same guarantees. We consider two offline methods: an exhaustive method for computing belief-based policies, and a heuristic search method for computing history-based policies.

DYNAMIC PROGRAMMING

The cost of reaching the goal from a belief state b following a policy π, denoted as $V^\pi(b)$, can be obtained as the solution of the Bellman equation

$$V^\pi(b) = \begin{cases} 0 & \text{if } b \text{ is a goal belief,} \\ c(a,b) + \max_o V^\pi(b_a^o) & \text{otherwise} \end{cases} \qquad (5.3)$$

where $a = \pi(b)$, o ranges over the observations that are possible in b_a, and $c(a,b)$ is the cost of doing action a in the belief state b, which in the worst case is:

$$c(a,b) = \max_{s\in b} c(a,s) \qquad (5.4)$$

A policy π is optimal if it minimizes the costs $V^\pi(b)$ over all beliefs b. The cost function V^π for an optimal policy $\pi = \pi^*$ is the optimal cost function V^*, that is the solution of Bellman's optimality equation

$$V(b) = \begin{cases} 0 & \text{if } b \text{ is a goal belief,} \\ \min_{a\in A(b)} [c(a,b) + \max_o V(b_a^o)] & \text{otherwise.} \end{cases} \qquad (5.5)$$

In the absence of *dead-ends* i.e., belief states b from which the goal cannot be reached, Equation 5.5 can be solved by a simple dynamic programming method called *Value Iteration* [Bellman, 1957], where the equation is used to update a value vector V over all beliefs b until a fixed point is reached. More precisely, in the version of Value Iteration (VI) known as Gauss-Seidel VI [Bertsekas, 1995], one starts with a value vector $V(b)$, initially set to 0 over all entries, and then iteratively updates each of the entries $V(b)$ over non-goal beliefs b as:

$$V(b) := \min_{a\in A(b)} [c(a,b) + \max_o V(b_a^o)]. \qquad (5.6)$$

In this setting, Value Iteration converges in a finite number of steps to the single solution $V = V^*$. The *optimal policy* π^* for solving the problem is then obtained from the *greedy policy* π_V

$$\pi_V(b) = \text{argmin}_{a\in A(b)} [c(a,b) + \max_o V(b_a^o)] \qquad (5.7)$$

using the value function $V = V^*$. We will see in Chapter 7 that the equations for solving POMDPs are similar but with the subexpression $\max_o V(b_a^o)$ representing cost in the worst case, replaced by the expected costs $\sum_o b_a(o)V(b_a^o)$. Likewise, the beliefs b, b_a, and b_a^o will become then probability distributions, and $b_a(o)$ will stand for the probability of observing o after doing the action a in b. A key difference when moving to POMDPs is that the set of possible beliefs b, representing probability distributions over the set of states, will no longer be finite.

HEURISTIC SEARCH: AO*

The problem with Value Iteration is that it is an exhaustive method that considers all belief states that are possible. While in the logical setting, this set is finite, it is still exponential in the number of states. We switch now to an alternative method that computes history-based policies incrementally. It is based on formulating these policies as solutions of an acyclic AND/OR graph that can be solved by the classical heuristic search algorithm AO* [Nilsson, 1980, Pearl, 1983].

An AND/OR graph is a rooted directed graph with three types of nodes: AND nodes, OR nodes, and terminal nodes. The terminal nodes can be either goal or failure nodes (dead-ends). An AND/OR graph is acyclic if there is no directed loop in the graph, i.e., no directed path that starts and ends in the same node. A solution to an acyclic AND/OR graph is a subgraph that includes the root node, one child of every OR node, all children of every AND node, and terminal nodes that are all goal nodes. In the absence of AND nodes, an AND/OR graph becomes a normal directed graph whose solution is a path from the root to a goal node. The childen of OR nodes represent choices, while the children of AND nodes represent contingencies, uncontrollable events, or adversarial moves that must all be handled in the solution of the problem. The cost of an AND/OR solution is the cost of its root node defined recursively as follows: the cost of terminal nodes is 0 and ∞ for goal and failure nodes respectively, the cost of AND nodes is the cost of the child with MAX cost, and the cost of OR nodes is the cost of the child with MIN cost plus one (assuming action costs equal to 1). Other cost structures are possible over AND/OR graphs as when the max operation associated with AND nodes is replaced by a sum or a weighted sum.

For capturing the history-based policies that solve a partially observable problem in terms of the solutions of an acyclic AND/OR graph, histories h are mapped to OR nodes $n(h)$, histories h extended with actions a applicable in h are mapped to AND nodes $n(h, a)$, and histories $\langle h, a, o \rangle$, where o is a possible observation after action a in history h, are mapped to the children of the AND node $n(h, a)$. The root node of the graph corresponds to the empty history, and the terminal goal nodes to the histories whose associated beliefs are goal beliefs. Solutions to such an implicit AND/OR graph must include the empty history h, and for every included non-goal history h, an action a, and all possible histories $\langle h, a, o \rangle$. The history-based policy encoded by such a solution is $\pi(h) = a$.

The algorithm AO* is a heuristic search method for solving acyclic AND/OR graphs. AO* maintains a graph G, called the *explicit graph*, that incrementally explicates part of the implicit AND/OR graph, and a second graph G^*, called the *best partial solution graph*, that represents an optimal solution of G under the assumption that the tip nodes n of G are terminal nodes whose values are given by a heuristic $h(n)$. Initially, G contains the root node of the implicit graph only, and G^* is G. Then, iteratively, a non-terminal tip node is selected from the best partial solution G^*, and the children of this node are explicated in G. The best partial solution G^* is then revised by a simple form of backward induction where the values of these new nodes are propagated up the (acyclic) graph. AO* finishes when the tip nodes of the best partial graph G^* are all terminal nodes. If the heuristic values are optimistic, the best partial solution G^* is then an optimal solution to the implicit AND/OR graph. If not, AO* produces a solution that is not necessarily optimal. Code for AO* is shown in Figure 5.1.

Informative heuristics are crucial for the performance of AO*. Heuristics for partially observable problems estimate the cost from a belief b to any goal belief. If the actions have deterministic effects, one of the approaches that have been used to obtain such estimates replaces the belief b by states s in b, resulting in *classical planning problems* whose cost can be estimated and combined in a number of ways [Bonet and Geffner, 2000, Bryce et al., 2006]. If estimates are admissible, the maximum over all such states s, yields an admissible heuristic, while the sum yields a non-admissible heuristic. These simplifications, however, remove the uncertainty in the belief b and hence ignore the need and value of sensing. Other heuristics for planning with sensing, called *cardinality heuristics*, focus exclusively on this uncertainty, defining the heuristic for b in terms of the number of states in b [Bertoli and Cimatti,

---

AO*

% G and G* are explicit and best graphs, initially empty; V′ is heuristic function.

Initialization

Insert node h_0 in G where h_0 is the empty history.
Initialize $V(h_0) := V'(b_0)$ where V' is admissible heuristic and b_0 initial belief.
Initialize best partial graph G^* to G.

Loop

Select non-terminal tip h from best partial graph G^*. If no such node, **Exit**.
Expand h in G: for each $a \in A(b)$ where b is belief associated with h, add node
   (h, a) as child of h, and for each observation o possible in b_a, add node (h, a, o)
   as child of (h, a). Initialize values $V(h, a, o)$ to the heuristic values $V'(b_a^o)$.
Update h and its ancestor AND and OR nodes in G, bottom-up as:

$$V(h, a) := 1 + \max_o V(h, a, o),$$
$$V(h) := \min_{a \in A(b)} V(h, a).$$

Mark best action in ancestor OR-nodes h to an action a with $V(h) = V(h, a)$,
   maintaining marked action if still best.
Recompute best partial graph G^* by following marked actions in G.

Figure 5.1: AO* for Computing History-based Policies for Partially Observable Problems.

2002]. More recent approaches have appealed to the translations developed for mapping conformant into classical planning problems (Section 4.2). Indeed, the delete-relaxation of a partially observable problem has a conformant solution once the preconditions of actions are pushed in as additional conditions of the actions' conditional effects [Hoffmann and Brafman, 2005]. Such relaxations have the advantage that they do not need to assume that the information is complete [Albore et al., 2009, 2011].

We have considered an exhaustive dynamic programming method (Value Iteration) for finding policies over a potentially cyclic belief space, and a heuristic search method (AO*) for finding policies over the acyclic space of histories. The two types of methods are not incompatible however. More recent algorithms like LAO* [Hansen and Zilberstein, 2001] and RTDP [Barto et al., 1995] manage to get the best of both worlds, and variations of these algorithms can be used to compute optimal and non-optimal belief-based policies in an incremental fashion using heuristics. We will consider such algorithms in the next chapter.

5.4 ONLINE SOLUTION METHODS

Offline solution methods have an inherent limitation: the size of the policies required for solving a problem may have exponential size. A practical alternative is to avoid computing a whole solution before acting, and to decide on the action to do in the current situation, to do it, to observe the results, and to iterate this loop until the goal is reached. This is the idea of *online planning*. In Section 2.4, we have seen different methods for online planning in the classical setting, from selecting the action a to do in a state s, *greedily*, by minimizing the expression $Q(a, s) = c(a, s) + h(s')$, where h is a heuristic function and s' is the state that follows s after the action a is done, to various *lookahead schemes* where the choice of the action a in s is based on a deeper exploration of the local space around s, to *learning schemes* such as LRTA* where the heuristic function is updated as the online search progresses.

The methods for selecting actions in the classical setting are all available in the partially observable setting but with two modifications: first, the local search is not over states s but over belief states b; second, the local space to search is not an OR graph but an AND/OR graph. Thus, the *greedy action* to do in a belief state b is the one that minimizes the expression $Q(a, b) = c(a, b) + \max_o V(b_a^o)$, where V is the heuristic over beliefs, and learning algorithms like LRTA* need to update the values $V(b)$ to $\min_{a \in A(b)} c(a, b) + \max_o h(b_a^o)$. Similarly, anytime optimal algorithms to be used in the finite-horizon version of the problem should not be based on A* but on AO* [Bonet and Geffner, 2012a].

Yet, some of the best current online partially observable planners for *deterministic* problems, follow a different strategy, where the action to be done next in a belief state b is selected by solving a *classical planning problem* obtained from a simplification of the problem [Bonet and Geffner, 2011, Brafman and Shani, 2012b]. The classical plan that is obtained is executed as long as the observations that are gathered do not refute the assumptions made in the simplification. When they do, an alternative, more informed relaxation is constructed and solved, leading to a new classical plan, and so on. These *replanning* approaches build on a formulation that extends the translation-based approach introduced for conformant planning [Palacios and Geffner, 2009] to the partially observable setting [Albore et al., 2009]. They can be shown to be complete, hence reaching the goal when possible, provided that the translation is complete and that the problem features no dead-ends. In the presence of dead-ends, these schemes can benefit from the use of lookahead schemes. Still, the completeness of these replanners in the absence of dead-ends is no small feat given that online planners that plan over finite horizons may fall into a loop. These replanners, on the other hand, exhibit an exploitation-exploration property that precludes such loops, where in every replanning episode they either reach the goal or learn that one of the assumptions made in the simplification is wrong. These replanners, however, are restricted to deterministic problems only.

5.5   BELIEF TRACKING: WIDTH AND COMPLEXITY

We have assumed so far that keeping track of beliefs is sufficiently simple. In the worst case, however, the computation of the belief state b_a^o that follows the belief b after the action a and observation o, is exponential in the number of state variables. One way to deal with beliefs b containing many states is by representing them by logical formulas whose satisfying assignments are precisely the states in b [Bertoli et al., 2001, Bryce et al., 2006, To et al., 2011], or by logical formulas that capture the state trajectories that are possible given the past actions and observations, and which can be queried by state-of-the-art solvers for determining whether certain formulas are true in b [Brafman and Shani, 2012b, Hoffmann and Brafman, 2006]. Indeed, a planner just needs to perform *two types of tests on beliefs* b, namely, whether a *goal* $X = x$ is true in b, and whether a *precondition* $X = x$ of an action a is true in b. The first test determines whether a given history has achieved the goal or needs to be extended, the second, which actions a can be used to extend it. The third required test, namely, whether an *observation* o is possible in b_a, follows from the other two: o is impossible in b_a if both $X = x$ and its negation hold in b_a^o, implying that b_a^o is empty. We will thus focus on a method for belief tracking that is targeted at these two queries, and which is not necessarily complete for other queries. Such an incompleteness, however, does not compromise the completeness of the resulting planners.

A key result is that it is possible to keep track of the beliefs necessary for determining the truth of action preconditions and goals in time and space that are exponential in a *width* parameter associated with the problem, which in many domains of interest is bounded and small. The result follows from the reductions of deterministic conformant problems into classical planning problems [Palacios and Geffner, 2009], and the extensions developed for handling partial observability [Albore et al., 2009] and non-deterministic actions [Bonet and Geffner, 2012b]. In this last formulation, the global beliefs b are factored into local beliefs b_X, for each variable X appearing in a precondition or goal. Keeping track of these local beliefs b_X is exponential in the number of state variables that are *relevant* to X. If this measure is called the width of X, $w(X)$, the *width* of the problem is the maximum $w(X)$ over the state variables X appearing in preconditions or goals.

The notion of *relevance* underlying this complexity bound is related to the notion of relevance in Bayesian networks [Pearl, 1988], while exploiting the structure of goals and action preconditions, and the information that certain variables will not be observed. More precisely, a variable X is defined as an *immediate cause* of a variable Y in a problem P iff $X \neq Y$, and either X occurs in the body C of a conditional effect $C \rightarrow E_1 | \cdots | E_n$ where Y occurs in a head E_i, $1 \leq i \leq n$, or X occurs in a formula $W_a(Y = y)$ where Y is an observable variable and $y \in D_Y$. Then X is said to be *causally relevant* to Y if $X = Y$, X is an immediate cause of Y, or X is causally relevant to a variable Z that is causally relevant to Y. Finally, X is *relevant* to Y if X is causally relevant to Y, Y is causally relevant to X and X is an observable variable, or X is relevant to a variable Z that is relevant to Y. The set of state variables Y that are relevant to X is called the *context* of X, which defines the set of variables that must be tracked concurrently with X in order to know the possible values for X after an execution. Moreover, this context $Ctx(X)$ defines a projected subproblem P_X that is like P but with the state variables limited to those in $Ctx(X)$. Belief tracking in P_X is thus exponential in the width $w(X)$ of X, while sound and complete for determining the values of X that are possible after an execution. The *factored belief tracking algorithm* that tracks the beliefs b_X over each of the projected problems P_X, for each precondition and goal variable X in the problem P, is complete for planning and runs in time and space that are exponential in the problem width, which may be much smaller than the number of variables in the problem.

As an illustration, consider the DET-Ring domain [Cimatti et al., 2004] depicted in Figure 5.2, where an agent can move forward or backward along a ring with n rooms. Each room has a window that can be opened, closed, or locked when closed. Initially, the status of the windows is not known, the agent does not know his initial location, and the goal is to have all windows locked. A plan for this deterministic conformant problem is to repeat n times the actions $(close, lock, fwd)$, skipping the last fwd action (alternatively, fwd can be replaced by the action bwd throughout). The state variables for the problem encode the agent location $Loc \in \{1, \ldots, n\}$, and the status of each window, $W(i) \in \{open, closed, locked\}$, $i = 1, \ldots, n$. The location variable Loc is (causally) relevant to each window variable $W(i)$, but no window variable $W(i)$ is relevant to Loc or to $W(k)$ for $k \neq i$. The largest contexts are thus for the window variables which have size 2. As a result, the width of the domain is 2, which is independent of the number of variables for the problem that grows linearly with

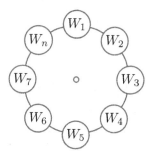

Figure 5.2: Ring problem with n windows that must be closed and locked. Initially, the agent does not know its location or the status of the windows. In NON-DET-Ring, each time the agent moves, the unlocked windows open or close non-deterministically. In another variation of the problem, the agent needs a key to lock the windows whose initial position is not known.

n. This means that belief tracking for this problem can be done in quadratic time since there are n contexts that need to be tracked, each of size $O(n)$ as the domain size for the window variables is constant. NON-DET-Ring is a variation of the domain where any movement of the agent, fwd or bwd, has a non-deterministic effect on the status of all windows that are not locked, capturing the possibility of external events that can open or close unlocked windows. This non-determinism has no effect on the relevance relation among the variables as Loc was already relevant to each variable $W(i)$. As a result, the change has no effect on the contexts or domain width that remains bounded and equal to 2. A further variation involves a key that is needed now to lock the windows, whose initial position is unknown. The agent may then perform a pick action that grabs the key when the key and the agent are in the same room. In all these variations, the problem width remains bounded and small. As a result, the *belief tracking task for planning* can be accomplished in low-order polynomial time even if the number and size of the beliefs is exponential in the number of rooms.

APPROXIMATIONS

Factored belief tracking is complete for planning and exponential in the problem width, yet this is still not good enough when problems have a large width. For such cases, however, it has been shown that it is possible to obtain meaningful approximations that are sound, polynomial, and powerful, even if not necessarily complete [Bonet and Geffner, 2013]. The idea is the consideration of a larger collection of projected subproblems P_X, each one involving a smaller set of variables. The algorithm being sound means that when a literal $X = x$ is reported as true or false after an execution, it is really true or false; while the algorithm being incomplete means that the literal $X = x$ may fail to be reported as true or false after an execution when $X = x$ is really true or false in the true belief. In the approximation, the variables X range not only over preconditions and goal variables, but also over *observable variables*, while the state variables that make it into the projected problem P_X are only those that are *causally relevant* to X. The result of this alternative decomposition is that there are more projected problems P_X but of smaller size whose local beliefs b_X can be tracked more efficiently. In this scheme, however, a state variable Y may be involved in two subproblems P_X and $P_{X'}$, such that $Y = y$ is known to be true in b_X but not known to be true in $b_{X'}$. The second step in the approximation is to enforce a local form of consistency among the local beliefs, an operation that can be achieved in polynomial time. The resulting approximation algorithm has been used successfully, in combination with simple heuristics, for solving large instances of domains like Minesweeper, Battleship, and Wumpus, where belief tracking is key [Bonet and Geffner, 2013].

5.6   STRONG VS. STRONG CYCLIC SOLUTIONS

A policy π that solves a partially observable problem P must be such that the only complete executions that are possible given π are *finite* and end up in a *goal belief*. There are problems, however, where these requirements are too strong and can't be ensured by any policy. For example, if the action of hammering a nail into a wall has cost 1 but fails half of the time, then the expected cost of getting the nail into the wall is 2 but the cost in the worst case is not bounded. The result is that there is no policy that solves the problem in the logical setting, although there is a perfectly good policy in the probabilistic setting. This policy does not guarantee that the problem will be solved in one step, but can guarantee that the goal will be solved eventually with probability 1. Interestingly, there is a way to weaken the requirements on policies in the logical setting so that they can capture exactly the policies that achieve the goal with probability 1. This requires the *assumption* that the non-deterministic state transition function $F(a, s)$ captures exactly the set of states s' that may follow action a in the state s with non-zero probability, even if the exact probabilities are not known. These solutions are called *proper policies* in the probabilistic setting [Bertsekas, 1995], and *strong cyclic policies* in the logical setting [Cimatti et al., 2003, Daniele et al., 1999]. The solutions that we have considered so far are called *strong policies*, as they solve the problem in a bounded number of steps. We characterize the strong cyclic policies below and review methods for computing them. Since the distinction between strong and strong cyclic policies is independent of issues pertaining to partial observability, we will assume for simplicity that the environment is *fully observable* and that the initial state s_0 is given.

In the fully observable setting, an execution or history is a sequence of states and actions $h = s_0, a_0, s_1, a_1, s_2, \ldots$. An *infinite execution* is *fair* when for each state-action pair s, a that appears an infinite number of times in the execution, the triplet s, a, s' also appears an infinite number of times for any $s' \in F(a, s)$. In other words, the infinite execution is *unfair* if the action a is applied an infinite number of times in state s, and yet there is a possible successor state s', $s' \in F(a, s)$, that only occurs a finite number of times.

Provided with this notion of fairness, we can define the *strong cyclic policies* as follows: π is a strong cyclic policy for P iff all the complete executions that are possible given π are either *finite and terminate in a goal state*, or are *infinite and unfair*. The difference with the *strong policies* for P is that infinite executions that do not reach a goal state are allowed as long as they are not fair. We will later discuss the equivalence between strong cyclic policies and proper policies for MDPs. We focus now on alternative methods for characterizing these policies and for computing them.

Since the cost in the worst case $V^{\pi}(s)$ associated with a strong cyclic policy π from a state s can be infinite, it is useful to introduce a second "optimistic" cost measure $V_{min}^{\pi}(s)$ that results from the assumption that the successor state s' of a non-deterministic action $a = \pi(s)$ in the state s is the "best" possible outcome for the agent. That is, while the cost in the *worst-case* function $V^{\pi}(s)$ is the solution to the equation:

$$V^{\pi}(s) = \begin{cases} 0 & \text{if } s \text{ is a goal state,} \\ c(a, s) + \max_{s' \in F(a,s)} V^{\pi}(s') & \text{otherwise for } a = \pi(s), \end{cases} \qquad (5.8)$$

which is analogous to Eq. 5.3 under the assumption that the observations are over full states, the "optimistic" cost function $V_{min}^{\pi}(s)$ is the solution to the equation:

$$V_{min}^{\pi}(s) = \begin{cases} 0 & \text{if } s \text{ is a goal state,} \\ c(a,s) + \min_{s' \in F(a,s)} V_{min}^{\pi}(s') & \text{otherwise for } a = \pi(s). \end{cases} \qquad (5.9)$$

That is, $V_{min}^{\pi}(s)$ measures the cost from s to the goal under the assumption that it is *the agent rather than nature* the one that chooses the successor state $s' \in F(a,s)$ that follows an action $a = \pi(s)$ in each state s. In particular, the cost $V_{min}^{\pi}(s)$ is finite when the agent can get from s to the goal following π if "lucky" enough, while $V_{min}^{\pi}(s)$ is infinite when no amount of luck would help the agent as there are no state trajectories linking s to the goal while following the policy π.

It turns out that π is a strong cyclic policy for a problem P with initial state s_0 iff the policy π is such that over all the states s that are reachable from s_0 following π, $V_{min}^{\pi}(s)$ is finite. The set of states reachable from s_0 and π is the minimal set of states S' that includes s_0 and any state s' such that $s' \in F(a,s)$ for $s \in S'$ and $a = \pi(s)$. In other words, π is strong cyclic when it drives the agent to states s all of which are separated from the goal by a finite trajectory $s_1, s_2, \ldots, s_n$ such that $s_1 = s$, the state s_n is a goal state, and $s_{i+1} \in F(a, s_i)$ for $a = \pi(s_i)$, $i = 1, \ldots, n-1$. It is easy to show indeed that infinite executions that feature a state s_i in the trajectory an infinite number of times, but do not feature the successor state s_{i+1} an infinite number of times, cannot be fair.

The simplification of the problem that underlies the optimistic cost function V_{min}^{π} has been used as a source of heuristics for non-deterministic and MDP problems where it is called the *min-min relaxation* [Bonet and Geffner, 2000, 2005]. It is also closely related to a different relaxation used in FF-Replan for solving MDPs, called the *deterministic relaxation* [Yoon et al., 2007]. Ignoring probabilities for the moment and focusing on semantics rather than in syntax, the min-min relaxation replaces each *non-deterministic action a* by *deterministic actions* $a_1, \ldots, a_m$, each one of which picks one of the possible outcomes of a, so that for any states s and s', $s' \in F(a,s)$ is true in the non-deterministic problem iff $s' = f(a_k, s)$ for some action a_k in the deterministic problem. When this relaxation is done at the syntactic level, it produces a *classical planning problem* where the uncertainty about non-deterministic transitions is now controlled by the agent. Indeed, it is easy to see that the cost function $V_{min}^{\pi}(s)$ is finite when such a classical planning problem has a solution from the state s.

This all suggests two methods for computing strong cyclic policies for non-deterministic but fully observable problems P. A purely semantic and exhaustive method is to compute first, via Value Iteration, the optimal cost function $V_{min}(s)$ for the min-min relaxation, where $V_{min}(s) = \min_{\pi} V_{min}^{\pi}(s)$. Then the states s for which $V_{min}(s) = \infty$ are removed from the problem, and the actions a that can possibly lead to such states from states s' are removed from the sets $A(s')$. This process of computing the value function V_{min} and pruning the action sets is iterated until the set of states s and the sets $A(s)$ of applicable actions do not change further.[1] If the initial state s_0 is removed in the process, the problem has no strong cyclic solution, else, the policy π that is greedy in the value function V_{min} computed last is one such solution [Daniele et al., 1999].

Alternatively, one can compute strong cyclic plans using *classical planners* over the deterministic relaxation P' of the problem P. Let $P'(s)$ be the classical problem obtained from P' by setting the initial situation to s. Define then a *complete state-plan (SP) pair* $\langle S', \Sigma \rangle$ as a set of states S', including

---

[1] An optimization is to select the states to prune from those which are reachable from the initial state s_0 with the policy π that is greedy in V_{min}. The iteration can be terminated when there are not such states.

the initial problem state s_0, along with a set Σ of classical plans $\sigma(s)$ for the problem $P'(s)$, one for each state $s \in S'$. The SP pair $\langle S', \Sigma \rangle$ is *consistent* when the plans $\sigma(s)$ in Σ that pass through a state $s' \in S'$ all apply the same action from P in s'. The expression $\Sigma(s')$ is used to denote this action. In particular, if the action is any of the determinizations a_k of a in P, $\Sigma(s)$ is a. The consistent SP pair $\langle S', \Sigma \rangle$ is *closed* when a state s' is in S' if there is state s in S' such that $s' \in F(a, s)$ for $a = \Sigma(s)$. It can then be shown that the partial policy π defined as $\pi(s) = \Sigma(s)$ for *complete SP pairs* $\langle S', \Sigma \rangle$ that are *consistent and closed*, is a *strong cyclic policy* for P. This means that a strong cyclic policy for P can be computed using classical planners incrementally, starting with the initial incomplete SP pair $\langle S', \Sigma \rangle$ where $S' = \{s_0\}$ and $\Sigma = \emptyset$. For this, classical plans are added to Σ to make the pair complete, and states are added to S' to make the pair closed. The state-plan pair is kept consistent by forcing the classical planner to respect the partial policy encoded by the pair. This can be achieved by adjusting the deterministic relaxations incrementally, or by modifying the classical planner used. An algorithm of this form will compute strong cyclic policies backtrack-free in problems with no dead-ends, but may have to backtrack otherwise. The first use of classical planners for computing strongly cyclic plans in this way is due to Kuter et al. [2008], and recent refinements to Fu et al. [2011] and Muise et al. [2012]. These are all offline algorithms. The planner FF-Replan mentioned above and to be discussed again in the next chapter, can be regarded as an online version of these algorithms.

CHAPTER 6

MDP Planning: Stochastic Actions and Full Feedback

Markov Decision Processes (MDPs) generalize the model underlying classical planning by allowing actions with stochastic effects and fully observable states. In this chapter, we look at a variety of MDP models and the basic algorithms for solving them: from offline methods based on dynamic programming and heuristic search, to online methods where the action to do next is obtained by solving simplifications, like finite-horizon versions of the problem or deterministic relaxations.

6.1 GOAL, SHORTEST-PATH, AND DISCOUNTED MODELS

There is a variety of MDP models, some more expressive than others [Bertsekas, 1995, Boutilier et al., 1999, Puterman, 1994]. We focus first on *Goal MDPs* that provide a direct generalization of the model underlying classical planning where the deterministic transition function $f(a, s)$ is replaced by transition probabilities $P_a(s'|s)$. At the same time, while the next state cannot be predicted with certainty, it is assumed to be *fully observed*. A Goal MDP is thus given by:

- a finite and non-empty state space S,

- an initial state $s_0 \in S$,

- a non-empty subset $S_G \subseteq S$ of goal states,

- sets $A(s)$ of actions applicable at each state $s \in S$,

- transition probabilities $P_a(s'|s)$ for s' being the next state after doing the action $a \in A(s)$ in the state $s \in S$, and

- positive action costs $c(a, s)$ for applying action $a \in A(s)$ in the state $s \in S$.

The planning task over Goal MDPs is to come up with an action strategy for reaching the goal with certainty given the uncertain effect of the actions and the observations gathered. The *solution form* for Goal MDPs cannot be thus a fixed action sequence as in classical planning; it must take observations into account. This is simple to do in MDPs, however, where observations are over full states and the dynamics and costs are Markovian, meaning that future states and costs depend on the current state but not on the previous history. The result is that the choice of the action to do next in MDPs just needs to take into account the last observation, and the solution form for MDPs is a function mapping (the observed) states into actions. These functions are called *closed-loop control policies* or simply *policies*, denoted by the symbol π. We will assume for now that a policy π maps every non-goal state s into an action $a \in A(s)$. Policies of this type are said to be *deterministic* and *stationary*. A *stochastic* policy π, on the other hand, is a function that maps states into probability distributions over actions, and a *non-stationary* policy is a function of both state and time. Stochastic and non-stationary policies can be used for controlling MDPs, but they are not strictly needed except in the setting of finite-horizon MDPs where optimal policies can be non-stationary.

A (deterministic and stationary) MDP policy π and state s define a *probability* for every state trajectory $\langle s_0, s_1, \ldots, s_{n+1} \rangle$ given by the product

$$P(s_0|s) \, P_{a_0}(s_1|s_0) \, P_{a_1}(s_2|s_1) \cdots P_{a_n}(s_{n+1}|s_n) \tag{6.1}$$

where $a_i = \pi(s_i)$ is the action dictated by the policy π in the state s_i, $P_{a_i}(s_{i+1}|s_i)$ is the state transition probability, and $P(s_0|s)$ is 1 if $s_0 = s$ and else is 0.

The *accumulated cost* of a state trajectory $\langle s_0, s_1, \ldots, s_{n+1} \rangle$ given a policy π, $a_i = \pi(s_i)$, is given in turn by the sum

$$c(a_0, s_0) + c(a_1, s_1) + \cdots + c(a_n, s_n). \tag{6.2}$$

The *expected cost* to reach the goal from state s using the policy π, denoted as $V^\pi(s)$, stands for the sum of the accumulated costs of the different state trajectories that are possible given π, weighted by their probabilities. The expected cost function V^π can also be characterized as the solution to a set of linear equations. For this, it is convenient to assume that goal states are *absorbing* and *cost-free*, meaning that some action a is applicable in each goal state s, and that such applicable actions a in a goal state s have zero costs and null effects; i.e., $c(a, s) = 0$ and $P_a(s|s) = 1$. Under the assumption that every policy π selects one of these "dummy" actions in each goal state, the expected cost of policy π from the state s, $V^\pi(s)$ can be defined by the expression

$$V^\pi(s) = E^\pi_s\left[\sum_{i \geq 0} c(\pi(X_i), X_i)\right] \tag{6.3}$$

where X_i is a random variable that represents the state at time i, and $E^\pi_s[\cdot]$ is the expectation with respect to the probability distribution on state trajectories that start in the state s given by (6.1). Moving the first term of the sum out of the expectation, the following fixed point equation is obtained

$$V^\pi(s) = c(\pi(s), s) + \sum_{s' \in S} P_{\pi(s)}(s'|s) V^\pi(s') \tag{6.4}$$

that defines the function V^π as the solution of a system of $|S|$ linear equations with the border condition $V^\pi(s) = 0$ for all goal states s.

It is possible to show that a policy π for a Goal MDP has a finite expected cost $V^\pi(s)$ if and only if starting in the state s, the application of the policy π leads to a goal state with probability 1. A policy π that leads to the goal with certainty for any possible initial state is called a *proper* policy. A necessary and sufficient condition for a policy π to be proper is that for any state s, there is a finite state trajectory $\langle s_0, s_1, \ldots, s_{n+1} \rangle$, starting in the state $s_0 = s$ and ending in a goal state s_{n+1}, such that all the state transitions in the trajectory are possible given π; i.e., $P_{\pi(s_i)}(s_{i+1}|s_i) > 0$ for $i = 0, \ldots, n$. Notice that the exact value of these probabilities does not matter as long as they are different than zero. This explains the correspondence between the proper policies in the probabilistic setting, and the strong cyclic policies in the non-deterministic setting analyzed in Section 5.6 that do not involve probabilities at all.

We will consider Goal MDPs where there are no *dead-ends*, i.e., states from which the goal cannot be reached. Formally, dead-ends are states s such that there is no state trajectory $\langle s_0, \ldots, s_{n+1}\rangle$ with $s_0 = s$, goal state s_{n+1}, and actions $a_0, \ldots, a_n$ such that the transition probabilities $P_{a_i}(s_{i+1}|s_i)$ are all positive for $i = 0, \ldots, n$. Clearly, if s is a dead-end, $V^\pi(s)$ is infinite for any policy π, and alternatively, if there are no dead-ends, there must be a policy π that is proper. We will relax the no dead-ends assumption for Goal MDPs when considering methods that compute *partial* policies.

A policy π is *optimal* for state s if $V^\pi(s)$ is minimum among all policies; i.e., $V^\pi(s) = \min_\pi V^\pi(s)$. While the optimal policies for Goal MDPs are the policies π that are optimal for the given initial state s_0, we will follow the standard notion that identifies the *optimal policies* as the policies that are optimal over all states. The cost function V^π for an optimal policy π is the optimal cost function V^*, which can be characterized as the unique solution of Bellman's optimality equation [Bellman, 1957]:

$$V(s) = \min_{a \in A(s)} [c(a,s) + \sum_{s' \in S} P_a(s'|s)V(s')] \tag{6.5}$$

for all non-goal states s, and $V(s) = 0$ for goal states. A deterministic, stationary *optimal policy* π^* can be obtained from the optimal cost function V^*, from the *greedy policy* π_V:

$$\pi_V(s) = \operatorname{argmin}_{a \in A(s)} [c(a,s) + \sum_{s' \in S} P_a(s'|s)V(s')] \tag{6.6}$$

with the value function V set to V^*. The ties in (6.6) can be broken arbitrarily.

Figure 6.1(a) depicts a simple example of a Goal MDP in which there is an agent that has to navigate in a grid with obstacles from the cell marked A to the cell marked G. The agent can move one cell at a time in each of the four directions as long as there are no obstacles, and the intended moves succeed with high probability while leaving the agent in nearby cells with non-zero probability. The panel (b) in Figure 6.1 shows a proper policy π for the problem as the action $\pi(s)$ to do at each of the cells s, except at the cell G representing the goal state.

SHORTEST-PATH AND DISCOUNTED MDPS

Stochastic Shortest-Path MDPs (SSPs) generalize Goal MDPs by dropping the requirement that action costs $c(a,s)$ over non-goal states s are positive [Bertsekas, 1995]. Instead, such action costs $c(a,s)$ can be either negative or zero, as long as any policy π that is not proper for a state s has an infinite expected cost $V^\pi(s)$. A policy is not proper for a state s when the probability of reaching the goal from s following the policy is less than 1. As for Goal MDPs, SSPs assume that there exists one policy that is proper for all the states.

Discounted Cost-based MDPs do not require the presence of absorbing and cost-free goal states, or the assumption that action costs are positive. Instead, discounted models assume that future costs depreciate over time according to a fixed rate $0 < \gamma < 1$, called the *discount factor*. In contrast to Eq. 6.2, the accumulated cost associated to a state trajectory $\langle s_0, \ldots, s_n\rangle$ under a policy π is

$$c(a_0, s_0) + \gamma\, c(a_1, s_1) + \cdots + \gamma^n\, c(a_n, s_n) \tag{6.7}$$

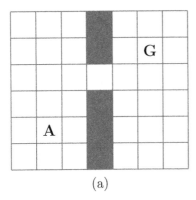

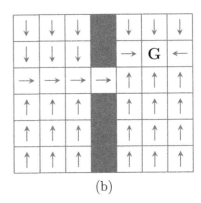

Figure 6.1: A Goal MDP in which an agent, initially at A, must reach the cell marked with G with certainty. The grey cells are obstacles that cannot be crossed. Each action moves the agent in the intended direction with non-zero probability but can also leave the agent in a nearby cell with non-zero probability as well. Panel (b) shows a proper policy for the problem depicted as the action to be done in each non-goal state.

where $a_i = \pi(s_i)$. Since the cost of any such trajectory is bounded from below and above by $\underline{c}/(1-\gamma)$ and $\bar{c}/(1-\gamma)$ respectively, where $\underline{c}$ and $\bar{c}$ are lower and upper bounds on the action costs $c(a, s)$, it follows that the expected costs $V^\pi(s)$ for all policies π and states s are finite. The Bellman equation characterizing this cost function is:

$$V^\pi(s) = c(\pi(s), s) + \gamma \sum_{s' \in S} P_{\pi(s)}(s'|s)V^\pi(s'), \tag{6.8}$$

and similarly, the optimal cost function V^* for Discounted Cost-based MDPs is given by the unique solution of the optimality equation [Bertsekas, 1995, Puterman, 1994]:

$$V(s) = \min_{a \in A(s)} [c(a, s) + \gamma \sum_{s' \in S} P_a(s'|s)V(s')]. \tag{6.9}$$

Discounted Reward-based MDPs are like Discounted *Cost-based* MDPs but with costs $c(a, s)$ replaced by rewards $r(a, s)$, and minimization of expected costs replaced by maximization of expected rewards. An example of a reward-based MDP is one where an agent gets a positive reward of R every time it reaches a piece of food, that once consumed appears randomly at a different location. A discount factor of $\gamma < 1$ ensures that the maximum discounted reward accumulated never exceeds $R/(1-\gamma)$.

The results and algorithms for Goal and Stochastic-Shortest Paths apply with small modification to Discounted MDPs. Moreover, Discounted MDPs can be easily compiled into *equivalent* Goal MDPs through a simple and efficient transformation [Bertsekas, 1995]. Thus, while certain problems, like the one above, can be more naturally expressed as Discounted MDPs, Discounted MDPs are not more expressive than Goal MDPs, and actually the opposite seems to be true as there is no known method for transforming general Goal MDPs into Discounted MDPs.[1]

---

[1]Some Goal MDPs can be transformed into equivalent Discounted MDPs, but the transformation is not general. For example, a Goal MDP with action costs $c(a, s)$ that are all uniform and equal to 1, can be transformed into an equivalent Discounted

In order to make precise the notion of *equivalence* among different types of MDPs [Bonet and Geffner, 2009], let us say that two MDPs M and M', possibly of different types, are *equivalent* iff they have the same set of non-goal states and actions (and hence the same space of policies), and for any policy π, the value functions V_M^π and $V_{M'}^\pi$ over M and M' are related by two constants α and β through the linear equation

$$V_{M'}^\pi(s) = \alpha V_M^\pi(s) + \beta \tag{6.10}$$

over all non-goal states s. The equation ensures that policies have the same relative ranking in M and M'; i.e., $V_M^\pi(s) < V_M^{\pi'}(s)$ iff $V_{M'}^\pi(s) < V_{M'}^{\pi'}(s)$. The constant α can't be zero, and is negative only when M and M' have different signs: one being cost-based and the other reward-based.

For showing that a Discounted Reward-based MDP M can be transformed into an equivalent Goal MDP M', one can show 1) that M is equivalent to a Discounted Reward-base MDP M_1 that is like M but with a negative constant R added to all rewards to make them all negative, 2) that M_1 is equivalent to a Discounted Cost-based MDP M_2 where these negative rewards are transformed into positive costs, and 3) that M_2 is equivalent to a Goal MDP M' that is like M_2 but with a new (absorbing, cost-free) goal state t added, such that the transition probabilities P' in M' are expressed in terms of the transition probabilities P of M, M_1, and M_2 as:

$$P_a'(s'|s) = \begin{cases} \gamma\, P_a(s'|s) & \text{if } s' \neq t \\ 1 - \gamma & \text{if } s' = t. \end{cases} \tag{6.11}$$

In this expression s ranges over the states in the original discounted model M, and a is an action applicable in s. Notice that in the resulting Goal MDP, every policy is proper, as every applicable action a in each non-goal state s, maps s into the goal state t with a non-zero probability $1 - \gamma$. The equivalence between the Discounted Reward MDP M and the Goal MDP M' then follows from the relation between the value functions V_M^π and $V_{M'}^\pi$ that satisfies (6.10) for $\alpha = -1$ and $\beta = -R/(1 - \gamma)$.

FINITE-HORIZON MDPS

The MDPs above are said to be *infinite horizon*, as the costs and rewards accumulate over a horizon that is not bounded a priori. Finite-horizon MDPs, on the other hand, are concerned with the accumulation of costs or rewards over a fixed number H of stages, called the problem *horizon*. Finite-horizon MDPs can be converted into infinite-horizon MDPs by simply augmenting the problem states s with the horizon left d. Thus, if s_0 is the initial state of the finite-horizon MDP M with horizon H, then the equivalent infinite-horizon MDP M' will have the pair $\langle s_0, H \rangle$ as the initial state, the pairs $\langle s, 0 \rangle$ as the goal states, and transition probabilities $P_a'(\langle s', d-1 \rangle | \langle s, d \rangle)$ equal to the transition probabilities $P_a(s'|s)$ in M. The same transformation is used for costs. The resulting Goal MDP M' has an important characteristic; namely, it is *acyclic*, meaning that the probability of any state trajectory starting and ending in the same state is zero. This is because time moves forward, and states $\langle s, d \rangle$ can only transition to states $\langle s', d-1 \rangle$, and this only when $d \neq 0$. Dynamic programming procedures like (Asynchronous) Value Iteration, to be considered next, can be used to solve finite-horizon MDPs,

---

Reward-based MDP with discount factor γ, $0 < \gamma < 1$, and rewards $r(a, s) = 0$ over non-goal states, and $r(a, s) = 1$ over goal states. The goal states remain absorbing but not cost-free in this Discounted MDP. This transformation, however, does not ensure equivalence when the action costs $c(a, s)$ are not uniform.

and more generally acyclic MDPs, very efficiently, in a single pass over all the states, by considering the states in order: first the states $\langle s, d \rangle$ with $d = 1$, then the states with $d = 2$, and so on, until reaching the states with $d = H$. Still, even a single-pass over all the states may not be computationally feasible. We will thus also consider incremental, heuristic search algorithms for solving finite-horizon MDPs and their use in online planning over general infinite-horizon MDPs (Section 6.4).

6.2   DYNAMIC PROGRAMMING ALGORITHMS

We focus next on the two standard dynamic programming algorithms for solving MDPs. While we focus on Goal MDPs, the methods apply to Stochastic Shortest-Path MDPs, and with minor mod-ification, to Discounted MDPs. We assume that there are *no dead-ends* and hence that the goal is reachable from all states. We will relax this assumption in the next section.

VALUE ITERATION

Value Iteration (VI) is a method for computing the optimal cost function V^*, which once plugged into the greedy policy π_V in place of V, yields the optimal policy π^*. The optimal cost function V^* is the unique solution to the optimality equation

$$V(s) = \begin{cases} 0 & \text{if } s \text{ is a goal state} \\ \min_{a \in A(s)} [c(a, s) + \sum_{s' \in S} P_a(s'|s)V(s')] & \text{otherwise.} \end{cases} \qquad (6.12)$$

Value iteration solves this equation by setting $V(s) = 0$ for goal states s and initializing the value of non-goal states arbitrarily, and then using Eq. 6.12 as an update

$$V(s) := \min_{a \in A(s)} [c(a, s) + \sum_{s' \in S} P_a(s'|s)V(s')] \qquad (6.13)$$

which is performed in parallel over all non-goal states. This operation, which is implemented by means of two value vectors V and V', is called a *full or parallel DP update*. Value iteration performs these paral-lel updates repeatedly. In the limit, the value vector V converges to the solution of Bellman's optimality equation (6.12), and hence to the optimal cost function V^*. Since the convergence is asymptotic, Value Iteration is stopped when the value vector V is such that the maximum difference between the expres-sions in the left and right-hand sides of (6.12) is small enough. If this difference, called the *residual* and defined as

$$Res_V \stackrel{\text{def}}{=} \min_{s \in S} |V(s) - [c(a, s) + \sum_{s' \in S} P_a(s'|s)V(s')]|, \qquad (6.14)$$

is sufficiently small, the policy π_V greedy with respect to V is optimal. More generally, the value of the residual can be used to bound the loss $V^{\pi_V}(s_0) - V^*(s_0)$ that results from following the greedy policy π_V from the initial state s_0 rather than an optimal policy. For Discounted MDPs, this loss is no greater than $2\epsilon\gamma/(1 - \gamma)$ if $Res_V < \epsilon$. While the loss can also be bounded in SSPs and Goal MDPs, the expression for the bound cannot be expressed in such a closed form [Bertsekas, 1995].

In a variation of VI, known as Asynchronous Value Iteration, the update in Eq. 6.13 is not performed over all states simultaneously but over some selected states. Provided that every state is updated infinitely often, Asynchronous VI also converges asymptotically to the optimal cost function

VALUE ITERATION

    Starts with value function stored in vector V with $V(s) = 0$ for goal states s

    repeat

        flag := true

        for each non-goal state s **do**

            new-value := $\min_{a \in A(s)} [c(a, s) + \sum_{s' \in S} P_a(s'|s) V(s')]$

            If $|V(s) - \text{new-value}| \geq \epsilon$ then

                $V(s) := \text{new-value}$

                flag := false

            end if

        end for

    until flag = true

Figure 6.2: Simple Version of Asynchronous Value Iteration implemented using single vector that outputs a value function V with residual Res_V smaller than the parameter ϵ, $\epsilon > 0$.

V^* [Bertsekas and Tsitsiklis, 1989]. This implies, among other things, that the simple variant of Value Iteration, implemented using a single value vector that is updated one state at a time, is a form of Asynchronous VI that also converges to V^*. This version of VI is known as Gauss-Seidel VI. Code for a version of VI that delivers a value function with residual less than a given ϵ is shown in Figure 6.2.

    A suitable version of Asynchronous Value Iteration can be used to solve *acyclic MDPs*, including *finite-horizon MDPs*, in one pass. Recall that an MDP is acyclic when all state trajectories that start and end in the same state have zero probability, and that finite-horizon MDPs can be cast as infinite-horizon and acyclic MDPs where the state is extended with the information of the horizon-to-go. All that is needed for solving acyclic MDPs in a single pass is to order the updates so that a state s is updated before a state s' when there is a state trajectory from s' to s in the MDP with positive probability. In the case of finite-horizon MDPs, it suffices to update states $\langle s, d-1 \rangle$ before states $\langle s', d \rangle$. It is simple to prove by induction that the values computed in one such pass are optimal. The procedure is also known as *backward induction* [Bertsekas, 1995].

POLICY ITERATION

The other standard dynamic programming algorithm for solving MDPs is Policy Iteration [Bertsekas, 1995, Howard, 1971, Puterman, 1994]. Whereas VI iterates over value functions in order to compute the optimal value function V^*, Policy Iteration (PI) iterates over policies, each one strictly better than the one before. Since the total number of (deterministic) policies is finite, PI converges to the optimal policy in a number of iterations that is bounded.

    Policy iteration applies two operations in sequence, starting with a *proper policy* $\pi = \pi_0$. First, it computes the value of the policy $V^\pi(s)$ over all states. It then finds a new policy π' that is proper and improves π if π is not optimal. The first step, called *Policy Evaluation*, is done by solving the set of $|S|$ linear equations given by (6.4) that characterize the value function V^π. The second step, called *Policy Improvement*, uses the value function V^π computed in the Policy Evaluation step, to see if there are states s where the actions dictated by the policy π are not best, under the assumption that after this first step, the policy π will be followed. For this, Q-factors of the form

$$Q^\pi(a, s) = c(a, s) + \sum_{s' \in S} P_a(s'|s)V^\pi(s') \tag{6.15}$$

are computed for each non-goal state s and action $a \in A(s)$. The policy π' that is like π except in states s for which the following strict inequality holds

$$\min_{a \in A(s)} Q^\pi(a, s) < Q^\pi(\pi(a), s) = V^\pi(s) \tag{6.16}$$

where $\pi'(s) = \mathrm{argmin}_{a \in A(s)} Q^\pi(a, s)$ is a policy that strictly improves π, i.e., $V^{\pi'}(s) \leq V^\pi(s)$ with the inequality being strict over some states. If there are no such states, π must be optimal, and Policy Iteration terminates.

Policy iteration produces a sequence of policies $\pi_0, \ldots, \pi_n$, starting with an initial proper policy π_0 such that each policy π_{i+1} is proper and strictly improves the previous one. The last policy π_n is a policy that cannot be improved further and is optimal. The length of the sequence is bounded by the total number of deterministic and stationary policies. A proper policy is needed initially in Policy Iteration, as the Policy Improvement step may not work when the expected costs are not bounded. For example, consider a Goal MDP with a single non-goal state s and two actions a and b such that a maps s into itself with probability 1, and b maps s into itself with probability $1/2$, and into the goal with probability $1/2$. If the initial policy π is such that $\pi(s) = a$, then $V^\pi(s)$ is infinite, and therefore both $Q^\pi(a, s)$ and $Q^\pi(b, s)$ are infinite as well, so that the policy $\pi'(s) = b$ does not appear to improve π even if π' is optimal and π is not. This can't happen, however, when π is a proper policy. Methods for computing proper or strong cyclic policies are discussed in Section 5.6. Yet, the *stochastic policy* that assigns to each state s an action in $A(s)$ with probability $1/|A(s)|$ is always proper in problems with no dead-ends. Code for a simple implementation of Policy Iteration is shown in Figure 6.3.

6.3   HEURISTIC SEARCH ALGORITHMS

Dynamic programming methods like VI and PI are exhaustive. They consider all the states in the problem from the very beginning, and thus cannot be used to solve problems with a large number of states; e.g., $|S| > 10^{10}$. *Heuristic search* methods, on the other hand, are incremental, and while they may end up considering many, if not all of the states in a problem, they use information about the *initial state* and lower bounds or *admissible heuristic functions*, to focus the search for solutions. In Sections 2.3 and 2.4, we reviewed heuristic search methods for solving directed (OR) graphs like A* and LRTA*, while in Section 5.3, we reviewed heuristic search methods for solving acyclic AND/OR graphs like AO*. In this section, we look at heuristic search algorithms for solving Goal MDPs, which can also be applied to Discounted MDPs after transforming them into Goal MDPs. One of the algorithms, RTDP, is a generalization of LRTA* to MDPs [Barto et al., 1995], another one, LAO*, is a generalization of AO* to *cyclic* AND/OR graphs. A crucial idea underlying these methods is that there is no need for computing *complete* policies that prescribe an action in every possible state when the initial state is given.

A *partial policy* π is a function that assigns an action $\pi(s)$ to some states but not necessarily to all of them. For a partial policy π to represent a solution to a Goal MDP, there is no need for π to be

POLICY ITERATION

    Starts with proper policy π and terminates with π being optimal

repeat

        Compute V^{π} and store it in vector V

        Let $\pi' := \pi$ be the new policy initialized to π

        Let change := false

        for each state $s \in S$ **do**

            for each action $a \in A(s)$ **do**

                $Q(a,s) := c(a,s) + \sum_{s' \in S} P_a(s'|s)V(s')$

            end for

            Let $Q := \min_{a \in A(s)} Q(a,s)$

            if $Q < V(s)$ **then**

                $\pi'(s) := \text{argmin}_{a \in A(s)} Q(a,s)$

                change := true

            end if

        end for

        if change = true **then** $\pi := \pi'$

until change = false

Figure 6.3: Policy Iteration for Goal MDPs. The algorithm terminates when π cannot be improved further and hence is optimal.

total, it suffices for π to be defined over the initial state of the MDP, s_0, and over the states s that can be reached from s_0 while following π. This set of states, denoted as $S(\pi, s_0)$, is the smallest set that includes s_0, and is closed in the following sense: if s is in $S(\pi, s_0)$, and $P_a(s'|s) > 0$ for $a = \pi(s)$, then s' must be in $S(\pi, s_0)$ as well. We say that such partial policies are *closed* with respect to s_0, or simply, that they are closed, leaving the initial state s_0 implicit. A state s is reachable from s_0 with policy π iff $s \in S(\pi, s_0)$.

    The main idea underlying heuristic search algorithms for MDPs can be expressed as follows. Let $Res_V(s)$ be the residual of the value function V over a non-goal state s, defined as the difference between the left and right-hand sides of Bellman's optimality equation at state s:

$$Res_V(s) = \left| V(s) - \min_{a \in A(s)} [c(a,s) + \sum_{s' \in S} P_a(s'|s)V(s')] \right|. \tag{6.17}$$

Bellman's result stating that V^* is the unique solution of the set of optimality equations and that the policy π_{V^*} greedy with respect to V^* is optimal, can be rephrased as saying that if V is a value function such that the residuals $Res_V(s)$ are zero over *all* the states, then π_V is optimal. Heuristic search algorithms for MDPs exploit a variation of this result that takes into account 1) the possibility of the value function V being an *admissible heuristic function*, i.e. a lower bound $V \leq V^*$, and 2) the fact that many states will not be reachable from the initial state s_0 and the greedy policy π_V.

FIND-AND-REVISE

Starts with value function V given by admissible heuristic function $h, h \leq V^*$

repeat

  Find one or more states s reachable from s_0 with π_V such that $Res_V(s) > \epsilon$

  Update V at those states s and possibly at other states

until no such states s found

Figure 6.4: Find-and-Revise: Computes value function V with residuals bounded by ϵ over all the states reachable with greedy policy π_V from the initial state s_0. For sufficiently small ϵ, the resulting greedy policy π_V is optimal for s_0.

In fact, provided that V is an *admissible heuristic*, it can be shown that the greedy policy π_V will be *optimal for the initial state s_0*, if the *residuals are zero over the states that can be reached with π_V from s_0* [Bonet and Geffner, 2003a]; i.e.,[2]

$$V^{\pi_V}(s_0) = V^*(s_0) \quad \text{if} \quad Res_V(s) = 0 \text{ for all } s \in S(\pi_V, s_0) \text{ and } V \leq V^*. \qquad (6.18)$$

This is a simple but important result that says that there is no need for the value function V to converge over all states for the greedy policy π_V to be optimal with respect to the initial state s_0; convergence over the states that are reachable from s_0 by following the policy π_V suffices. For this, the value function V must be admissible, else like in A*, the optimal solutions may be missed. As an example of this, consider a deterministic Goal MDP with two non-goal states s_0 and s_1 such that action a maps s_0 into the goal at cost 5, while b maps s_0 into s_1, and s_1 into the goal, in both cases at cost 1. In this problem, b is a better action than a in the state s_0. Yet, if the value function V is not admissible and overestimates the value of the state s_1, e.g., by making $V(s_1) = 10$, the policy π_V greedy in V will pick the action a rather than b in s_0, while the residuals $Res_V(s)$ will be zero over all the non-goal states s that are reachable from s_0 with the policy π_V. Actually, in this deterministic MDP, the same suboptimal solution would result from A* with the same non-admissible heuristic function. Thus, the admissibility of V in (6.18) is a necessary condition for the optimality of the policy π_V with respect to the initial state s_0.

The principle expressed in (6.18) suggests a simple generic method for computing admissible value functions V such that the residuals $Res_V(s)$ over the states s reachable from the initial state S_0 and the greedy policy π_V do not exceed a certain threshold $\epsilon > 0$. The generic method, shown in Figure 6.4, is called Find-and-Revise [Bonet and Geffner, 2003a]. Find-and-Revise proceeds in two stages: it first searches for one or more states s in $S(\pi_V, s_0)$ with residuals $Res_V(s) > \epsilon$, and then updates such states as in Asynchronous Value Iteration. This process is repeated until there are no more such states. If the initial value function $V = h$ is *monotonic*, the process terminates in at most $\frac{1}{\epsilon} \sum_{s \in S} V^*(s) - h(s)$ iterations.

The notion of *monotonicity* or *consistency* is well known in heuristic search over directed graphs, where it refers to heuristics h that satisfy the triangular inequality $h(n') \leq c(n, n') + h(n)$ for every

---

[2]We assume that a value function V determines a unique greedy policy π_V. This is easy to enforce by assuming a static ordering among actions so that ties in the choice of the action $\pi_V(s)$ are broken by using this ordering, e.g., preferring action a to b if a precedes b in the ordering.

LRTA*

% *Initial value function V given by admissible heuristic h*
% *Changes to V stored in hash table*

Let $s := s_0$
While s is not a goal state **do**
  Evaluate each action $a \in A(s)$ as: $Q(a, s) := c(a, s) + V(s')$ where $s' = f(a, s)$
  Select best action $\mathbf{a} := \operatorname{argmin}_{a \in A(s)} Q(a, s)$
  Update value $V(s) := Q(\mathbf{a}, s)$
  Set $s := f(\mathbf{a}, s)$
end while

Figure 6.5: Trial of Learning Real Time A*: LRTA* can be seen as instance of Find-and-Revise for deterministic MDPs and $\epsilon = 0$. LRTA* would still converge to an optimal solution as an instance of Find-and-Revise if trials were interrupted right after the first update that changes the value function.

pair of nodes n and n' that are connected through an edge with cost $c(n, n')$. The use of a monotonic heuristic in an algorithm like A* ensures that the evaluation function $f(n)$ of the nodes selected for expansion never decreases. In the probabilistic setting, the monotonicity of V translates into the inequality $V(s) \leq c(a, s) + \sum_{s' \in S} P_a(s'|s)V(s')$ for each state $s \in S$ and action $a \in A(s)$. Since updates preserve monotonicity (and admissibility), an initial monotonic value function guarantees that state values $V(s)$ never decrease after updates. Also, since these values are bounded from above by the finite optimal costs $V^*(s)$, they cannot change more than $(V^*(s) - h(s))/\epsilon$ times, where h is the initial admissible and monotonic value function.

In the implementation of Find-and-Revise procedures, the initial value function V is usually given by a heuristic function h, and changes in the value function are stored in a hash table. The search step in Find-and-Revise can be implemented in $O(|S|)$ time with a standard depth-first traversal that keeps track of visited nodes. A version of Find-and-Revise that combines such a depth-first search with a labeling procedure for marking states s as solved when the residual over all the states that are reachable from s and the greedy policy π_V is bounded by ϵ, is known as HDP for heuristic dynamic programming [Bonet and Geffner, 2003a]. We consider other Find-and-Revise variants below, starting with the LRTA* algorithm for deterministic problems presented in Section 2.4.

LRTA*

Learning Real Time A* (LRTA*) is a simple but powerful online search algorithm for finding paths in graphs [Korf, 1990]. As explained in Section 2.4, LRTA* evaluates the actions a applicable in the current state s, starting with the initial state s_0, by computing the factors $Q(a, s) = c(a, s) + V(s')$, where V is a value function initialized to a given heuristic h and s' is the successor state $s' = f(a, s)$. LRTA* chooses then the action a that minimizes these $Q(a, s)$ values, revises the value $V(s)$ to $Q(a, s)$, and iterates in this way by setting s to s' until reaching a goal state (Figure 6.5).

LRTA* can be seen as an Asynchronous Value Iteration algorithm over a *deterministic* MDP model, where the states that are updated are obtained by running simulations from the initial state,

RTDP

% *Heuristic h is the initial value function V*
% *Changes to V stored in a hash table*

Let $s := s_0$
While s is not a goal state **do**
  Evaluate each action $a \in A(s)$ as: $Q(a, s) := c(a, s) + \sum_{s' \in S} P_a(s'|s)V(s')$
  Select best action $\mathbf{a} := \text{argmin}_{a \in A(s)} Q(a, s)$
  Update value $V(s) := Q(\mathbf{a}, s)$
  Select next state s' with probability $P_{\mathbf{a}}(s'|s)$ and set $s := s'$
end while

Figure 6.6: Trial of RTDP: RTDP generalizes LRTA* to MDPs by evaluating actions using the Q-factors corresponding to MDPs, and by sampling the successor states.

selecting in each state s the action $\pi_V(s)$ that is greedy in V. Yet, Asynchronous Value Iteration is an exhaustive algorithm, while LRTA* is not. Indeed, LRTA* can converge to the optimal (partial) policy π_V without visiting most of the states in the problem. This is because LRTA* is not only an instance of Asynchronous VI, but of the general Find-and-Revise schema, and it thus exploits property (6.18), taking advantage of both a known initial state s_0 and an initial admissible value function. Actually, LRTA* would still converge to an optimal solution if trials were interrupted right after the first update that changes the value of a state. The resulting algorithm would be a version of Find-and-Revise with $\epsilon = 0$, where a single state is revised in each iteration, found by executing the greedy policy π_V from the initial state. The extra work done in each single trial by LRTA* is thus not necessary for convergence. Moreover, in the presence of dead-ends in the problem, LRTA* can be trapped into a dead-end, while this variant does not, if the problem has a solution.

REAL-TIME DYNAMIC PROGRAMMING

Real Time Dynamic Programming (RTDP) is a generalization of LRTA* for general Goal MDPs [Barto et al., 1995] that inherits its two main properties: in the absence of dead-ends, each trial eventually reaches the goal, and successive trials eventually converge to a value function V such that the policy π_V greedy in V is optimal for the initial state s_0. For this, two changes are done to LRTA*. First, the Q-factors for evaluating a in the state s are computed using the expression corresponding to the Bellman equation for MDPs; i.e., $Q(a, s) = c(a, s) + \sum_{s' \in S} P_a(s'|s)V(s')$. Second, in the simulated executions of the policy π_V, the state s' that follows action a in the state s is *sampled* with probability $P_a(s'|s)$. The resulting RTDP algorithm is shown in Figure 6.6.

By sampling successor states stochastically, RTDP converges more quickly on the states that matter the most, but less quickly on the states that are encountered with low probability. Indeed, if the value function V is such that there is a single state s with a residual greater than ϵ that is reachable with the greedy policy π_V from s_0, RTDP does *not* guarantee that this state will be found and updated in $O(|S|)$ time. This may actually take a large and unbounded number of trials if the probability of reaching s from s_0 while following the policy π_V is small. The algorithm Labeled RTDP is a modification of the RTDP algorithm that improves its convergence by ensuring that at least one such state is found in every RTDP trial when there are such states. If not, Labeled RTDP terminates [Bonet and Geffner, 2003b]. This is achieved, as in the algorithm HDP mentioned above, by labeling states s as solved when all the states s' that are reachable with the policy π_V from s have residuals $Res_V(s')$ no greater than ϵ. LRTDP terminates when the initial state s_0 is solved. Since the MDP can contain cycles, the labeling procedure is not done bottom-up, from children to parents, but using a version of Tarjan's algorithm [Tarjan, 1972] for detecting and labeling strongly connected components [Bonet and Geffner, 2003a,b].

LAO*

In Section 5.3 we considered the AO* algorithm for solving *acyclic* AND/OR graphs [Nilsson, 1980]. LAO* [Hansen and Zilberstein, 2001] is an extension of AO* for solving *cyclic* AND/OR graphs, whose solutions may contain cycles as well. Since Goal MDPs can be cast as cyclic AND/OR graphs whose (optimal) solutions encode the (optimal) solutions for Goal MDPs, LAO* can be used for solving Goal MDPs, and hence Discounted MDPs. In the AND/OR graph corresponding to a Goal MDP, the OR nodes correspond to the non-goal states s, the AND nodes correspond to the state-action pairs (s, a) where a is an action applicable in s, and the terminal nodes are goal states, and states from which no action can be applied. The children of the OR node s are the AND nodes (s, a), and the children of the AND node (s, a) are terminal or OR nodes s' for which $P_a(s'|s) > 0$.

AO* explicates the implicit AND/OR graph incrementally, starting with the graph G that contains just the root node, and maintains the subgraph G^* of G that encodes the optimal solution of G under the assumption that the nodes in G that have not yet been explicated (its children added to G), are terminal nodes n with values $V(n)$ given by a heuristic $h(n)$. AO* then proceeds to pick one of these tip nodes of G^* that are not terminal nodes of the original problem, and expands it in G, updating both G and its best solution subgraph G^*. AO* terminates when the tip nodes in G^* are all terminal nodes. The solution to the AND/OR graph is given then by G^*. The solution is optimal if the heuristic h is admissible.

The best solution subgraph G^* is obtained from G incrementally by propagating the values of the last children added to G to its parents and ancestors. This propagation is a single pass of Value Iteration (backward induction) that takes advantage of the acyclic structure of the AND/OR graph. If the acyclic AND/OR graph represents an acyclic MDP, the value $Q(a, s)$ of an AND node (s, a) is the function $Q(a, s) = c(a, s) + \sum_{s'} P_a(s'|s)V(s')$ of its children s', while the value of an OR node $V(s)$ is the minimum value $Q(a, s)$ of its children. The best solution subgraph G^* is updated by picking the best child (a, s) of each OR node s encountered during the bottom-up propagation of values.

In the presence of *cycles*, this method for maintaining the best subgraph G^* of G is no longer correct. Indeed, if the AND/OR graph represents a cyclic MDP, a single pass of Value Iteration over the ancestors of the nodes last added to the explicit graph G does not necessarily produce optimal values. Instead, Value Iteration must be run until the residuals over such nodes are smaller than a given ϵ parameter. This is precisely what LAO* does [Hansen and Zilberstein, 2001]. Since running Value Iteration until convergence in each expansion step of LAO* is too time consuming, an alternative to LAO*, called Improved LAO* or simply ILAO*, is normally used instead. The two key changes from LAO* to ILAO* are that ILAO* expands all non-terminal tip nodes of G^* in each step, and that the values are propagated up from the new nodes by updating the node values once. Conveniently, both operations can be done with a single depth-first traversal of the subgraph G^* in which the updates are done during backtracks. The result is that the resulting values are no longer optimal over G, and hence that G^* is not necessarily a best solution subgraph of G (so ILAO* is not a best-first algorithm like AO* or LAO*). Thus, when the termination condition of AO* and LAO* is reached in ILAO*, and G^* contains no tip that can be expanded (because such tips are terminal nodes), ILAO* runs Value Iteration until the residuals over the nodes in G^* are smaller than ϵ. If the optimal subgraph G^* does not change as a result, ILAO* terminates, else ILAO* continues using the revised subgraph G^*. ILAO* is not an exact instance of Find-and-Revise as the expansion of tip nodes of G^* does not necessarily imply that their residuals exceed ϵ (although this will often be the case). Still at termination, the states in G^*, which are those reachable with the greedy policy π_V from s_0, have all residuals that are bounded by ϵ. Further details on heuristic search algorithms for MDPs can be found in the book by Mausam and Kolobov [2012].

6.4   ONLINE MDP PLANNING

Online planning methods are not aimed at computing partial or complete policies, but at the selection of the action to do next in a planning-and-execution cycle. While they don't offer the guarantees of offline methods, they are usually informed by offline methods, and are more practical as they can be used over larger problems. In Sections 2.4 and 5.4, we reviewed online planning methods for classical and partially observable problems. All these methods can be adapted to the context of MDPs: greedy action selection using a *heuristic function* to estimate expected cost-to-go, less greedy selection methods following an exhaustive *lookahead* up to a certain depth H, *adaptive greedy action selection* following a number of iterations of LRTA* or RTDP that combine simulations and updates, as well as choices based on the solution of suitable problem *relaxations*. These various approaches are not incompatible with each other and admit a number of variations. Below we focus on two relaxation-based methods for online MDP planning: one that relies on the use of *classical planners over deterministic relaxations* of the MDP, and one that relies on *finite-horizon relaxations solved by anytime optimal methods*. Anytime optimal methods solve the problem incrementally, producing optimal solutions if given sufficient time, and (hopefully) good solutions when given shorter time windows.

CLASSICAL REPLANNING FOR MDPS

In Section 5.4 we considered the use of classical planners for computing strong cyclic solutions to fully observable non-deterministic problems. This same idea can be used for MDPs, by just replacing the transition probabilities P in the MDP by a non-deterministic state transition function F such that $s \in F(a, s)$ iff a is an action applicable in the state s and $P_a(s'|s) > 0$. The (possibly partial) strong cyclic policies for the resulting non-deterministic problem capture exactly the (possibly partial) proper policies for the MDP. The strong cyclic policies are computed by running a classical planner multiple times over a *deterministic relaxation* where each non-deterministic action a is replaced by a set of deterministic actions $a_1, \ldots, a_m$ with the same preconditions as a, each of which captures a possible effect of a; i.e., for any state s where action a is applicable, $s' \in F(a, s)$ iff $s' = f(a_i, s)$ for some $i \in \{1, \ldots, m\}$. This *deterministic relaxation*, which is implemented at the level of the compact representation of non-deterministic models and MDPs, can be used for online planning in those settings too. For this, in a state s, a classical plan is obtained using a classical planner over the compact representation of the deterministic relaxation, and the plan is executed until a state s' is observed that does not agree with the state predicted by the plan in the relaxation. In such a case, a classical planner is invoked again on the deterministic relaxation with s' as the initial state, and the process is repeated until a goal state is reached. This classical replanning approach to MDPs that selects actions ignoring the non-determinism and the actual probabilities (except for whether they are 0 or not) works surprisingly well in many domains, as shown by the online MDP planner FF-Replan [Yoon et al., 2007]. This basic approach can be improved in a number of ways such as mapping probabilities into costs, and using then a classical planner for approximating the minimum cost plans, which would then encode the most likely plans in the relaxation [Keyder and Geffner, 2008b, Mausam and Kolobov, 2012]. Alternatively, the resulting action selection mechanism can be used as a suitable *base policy* for seeding algorithms like UCT.

ANYTIME ALGORITHMS FOR FINITE-HORIZON RELAXATION

A second type of simplification used for infinite-horizon MDP planning is the finite-horizon relaxation. For a given state s and horizon H, this relaxation defines a finite-horizon MDP with initial state s, with the same transition probabilities and costs. This finite-horizon MDP represents a *different* infinite-horizon MDP that is acyclic, whose states s are augmented with information about the horizon-to-go d, and whose goals are extended to include the terminal states (s, d) for $d = 0$. Optimal or approximate algorithms can then be used to compute a policy for this relaxed MDP. The action dictated by this policy in the current state s is then applied, the resulting state is observed, and the process is repeated until a goal state of the original MDP is reached. However, rather than using small horizons H that result in finite-horizon MDPs that can be solved optimally in real-time, it has been found useful to consider larger horizons in combination with *anytime optimal methods*. We consider two such methods below. For the use of RTDP in an online MDP planning setting; see Kolobov et al. [2012a,b].

UCT

UCT is a Monte-Carlo Tree Search (MCTS) algorithm for solving finite-horizon MDPs and, more generally, AND/OR trees [Chaslot et al., 2008, Kocsis and Szepesvári, 2006]. UCT has been successfully used in a number of settings, including the game of Go [Gelly and Silver, 2007], real-time strategy games [Balla and Fern, 2009], and general game playing [Finnsson and Björnsson, 2008]. Like the standard heuristic search algorithm AO* for acyclic AND/OR graphs, UCT builds an explicit graph G incrementally. There are however four main differences between UCT and AO*. First, UCT selects the tip node to expand in G by running a *simulation* from the root node, which may add at most one new node to G. Second, UCT evaluates tip nodes by simulating a given *base policy* from the node. Third, values are propagated up the tree by means of *Monte-Carlo updates*. Four, UCT has *no termination* condition and its optimality over finite-horizon MDPs is only asymptotic. We explain these aspects of UCT in more detail below. Code for UCT is shown in Figure 6.7.

   UCT consists of a sequence of *stochastic simulations* that start at the root node of the AND/OR tree for the finite-horizon MDP. When this simulation reaches a node that is not in the explicit graph, the node is added to the graph, and the heuristic value of the node is obtained by executing a given *base policy* from the node. The processing done by UCT is aimed at improving the quality of this base policy at the root node. While the simulation traverses internal nodes of the explicit graph, the successor states are sampled stochastically, as in RTDP, but the choice of the actions is not greedy on the Q-values, but on the sum of the Q-values plus a *bonus term* equal to

$$- C \sqrt{2 \log N(s,d)/N(a,s,d)} \qquad (6.19)$$

that ensures that all the applicable actions would be tried infinitely often at suitable rates. In Eq. 6.19, C is an exploration constant, and $N(s,d)$ and $N(a,s,d)$ are counters that track the number of simulations that have passed through the node (s,d) in the tree, and the number of times that action a has been selected at such node. If a has never been tried at s, $N(a,s,d) = 0$ and the bonus term is $-\infty$, forcing a to be selected unless there are other unexplored actions. The bonus term is based on a similar term used in UCB [Auer et al., 2002], a regret-optimal algorithm for the multi-armed bandit problem.

The counters $N(s,d)$ and $N(a,s,d)$ are maintained for the nodes in the explicit graph only. When a node (s,d) is generated that is not in the explicit graph, the node is added to the explicit graph, the counters $N(s,d)$, $N(a,s,d)$, and $Q(a,s,d)$ are allocated and initialized to 0, and a cost $c(\pi, s, d)$ is sampled by simulating the base policy π for $H - d$ steps starting at s, and propagating this sampled cost upward along the nodes in the simulated path. These values are not propagated using full Bellman backups as in AO*, RTDP or VI, but through Monte-Carlo backups that update the current average with a new sampled value. For a successful use of a UCT-like algorithm in domain-independent MDP planning, see Keller and Eyerich [2012].

ANYTIME AO*

Anytime AO* is a simple variation of the AO* algorithm for AND/OR trees aimed at bridging the gap between AO* and UCT [Bonet and Geffner, 2012a]. Like UCT and unlike AO*, Anytime AO* is anytime optimal even in the presence of *non-admissible* or *random heuristics*. A random heuristic is a heuristic that corresponds to a random variable that can be sampled such as the *cost of a base policy*. On the other hand, Anytime AO* like AO* and unlike UCT has a clear termination condition that is achieved when there are no more nodes to add to the explicit graph.

Anytime AO* is the result of two small changes in AO*. The first, designed to handle non-admissible heuristics, is that rather than always selecting non-terminal tip nodes from the explicit graph G that are part of the best partial solution graph G^*, Anytime AO* selects non-terminal tip nodes from $G \setminus G^*$ with some positive probability. The second change, designed for dealing with random heuristics h, is that when the value $V(s,d)$ of a tip node (s,d) is set to a heuristic $h(s,d)$ that is a *random variable*, such as the cost obtained by following a base policy π for d steps from s, Anytime AO* uses *samples* of $h(s,d)$ until the node (s,d) is expanded. Until then, each time the value $V(s,d)$ is required, which occurs each time that a parent node of (s,d) is updated, a new sample of $h(s,d)$ is obtained which is averaged with the previous samples. This is implemented in standard fashion by incrementally updating the value $V(s,d)$ using a counter $N(s,d)$ and the new sample.

Anytime AO* has been tried as an online planning algorithm over different types of MDPs, including problems like the Canadian Traveller Problem where it appears to do as well as UCT, which represents the state-of-the-art [Eyerich et al., 2010]. An advantage of Anytime AO* is that it can potentially benefit from a number of techniques developed for speeding up A* and AO* like the use of weights $W > 1$ in the heuristic term [Chakrabarti et al., 1988]. Two advantages of UCT, on the other hand, are that it is a *model-free* method that can work perfectly well with a simulator rather than a model, and that it is less affected by large *branching factors* that obtain when the number of states s' that may result from doing an action a in a state s is large.

UCT(s, d)

    if $d = 0$ or s is goal **then return** 0

    if (s, d) is not in explicit graph G **then**

        Add node (s, d) to explicit graph G
        Initialize $N(s, d) := 0$ and $N(a, s, d) := 0$ for all $a \in A(s)$
        Obtain sampled accumulated cost $c(\pi, s, d)$ by simulating base policy π for
        $H - d$ steps starting at s
        return $c(\pi, s, d)$

    if node (s, d) is in explicit graph G **then**

        $Bonus(a) := C \sqrt{2 \log N(s, d) / N(a, s, d)}$ if $N(a, s, d) > 0$, else ∞, for $a \in A(s)$
        $a^* := \operatorname{argmin}_{a \in A(s)}[Q(a, s, d) - Bonus(a)]$
        Sample state s' with probability $P_a(s'|s)$
        Let $Cost := c(a, s) + \text{UCT}(s', d - 1)$
        Increment $N(s, d)$ and $N(a, s, d)$
        Set $Q(a, s, d) := Q(a, s, d) + [Cost - Q(a, s, d)]/N(a, s, d)$
        return $Cost$

Figure 6.7: UCT for finite-horizon cost-based MDPs: H is the horizon, G is the explicit graph (initially empty), π is the base policy, and C is the exploration constant. Procedure is called over node (s, H) where s is the current state. When time runs out, UCT selects the action applicable at s that minimizes $Q(a, s, H)$.

6.5 REINFORCEMENT LEARNING, MODEL-BASED RL, AND PLANNING

Reinforcement learning methods are algorithms that learn a control policy by trial-and-error through a process that seeks to maximize expected rewards [Sutton and Barto, 1998]. Modern reinforcement learning algorithms can actually be understood as solving a Discounted Reward-based MDP whose probabilistic and reward parameters are not known to the agent, who can nevertheless interact with a real or simulated world that is governed by such an MDP. Q-learning is one of the simplest such algorithms [Watkins, 1989]. Starting with arbitrary $Q(a, s)$ values, and given an execution $s_0, a_0, r_0, s_1, a_1, r_1, \ldots$ where s_i, a_i, and r_i represent the states, actions, and rewards received at time i, Q-learning updates the value $Q(a_i, s_i)$ of action a_i in the state s_i after observing the next state s_{i+1} and getting the reward r_{i+1} through the expression:

$$Q(a_i, s_i) := (1 - \alpha_i) \, Q(a_i, s_i) + \alpha_i \left(r_{i+1} + \gamma \max_{a \in A(s_{i+1})} Q(a, s_{i+1}) \right) \tag{6.20}$$

where γ stands for the discount factor and the α_i parameters represent the learning rate. The term

$$r_{i+1} + \gamma \max_{a \in A(s_{i+1})} Q(a, s_{i+1}) \tag{6.21}$$

can be understood as a *stochastic sample* of the term

$$\sum_{s \in S} P_{a_i}(s|s_i) \left[r(s_i, a_i, s) + \gamma \max_{a \in A(s)} Q(a, s) \right],$$ (6.22)

which would be used to update the $Q(a_i, s_i)$ factor if the probability and reward parameters were known.[3] The reason for learning $Q(a, s)$ values as opposed to $V(s)$ values, is that the latter cannot be used for selecting actions without knowing the probabilities and rewards. The key result for this type of stochastic updates is that they converge to the optimal Q-values in the limit when all actions are tried in all states sufficiently often, provided that the constants α_i comply with basic requirements [Bertsekas and Tsitsiklis, 1996, Sutton and Barto, 1998, Szepesvári, 2010]. In Q-learning this convergence is achieved by choosing a random action in each state s with small but non-zero probability ϵ, choosing otherwise the greedy action a, i.e., the action that appears to be best according to the current Q-values, namely, $a = \text{argmax}_{a \in A(s)} Q(a, s)$.

Q-learning is a *model-free* algorithm for solving MDPs as it learns the behavior but not the model (parameters). *Model-based reinforcement learning* algorithms, on the other hand, are aimed at learning both the model and the behavior, which they derive from the model like any planning-based method. Interestingly, some of the best known model-based RL algorithms like R-MAX [Brafman and Tennenholtz, 2003] actually map the learning problem into a planning problem. Indeed, R-MAX plans in an *optimistic model* where rewards $r(s, a, s')$ that are not yet known are replaced by known, optimistic rewards R, $r(s, a, s') \leq R$, and similarly, transition probabilities $P_a(s'|s)$ that are not yet known are replaced by known transition probabilities $P_a(s_R|s) = 1$, where s_R is a new and absorbing "nirvana" state where the rewards $r(s_R, a, s_R)$ are all equal to the upper bound R. Planning in this "optimistic" model directs the learning agent to states s and actions a that lead to the "nirvana" state s_R. By repeating this process over and over, a sufficient number of samples is obtained for the unknown parameters $P_a(s'|s)$ and $r(s, a, s')$ until they become known with sufficient confidence. When this happens, the optimistic values for these parameters are replaced by the learned values, and the optimistic MDP model becomes less optimistic and more accurate. By iterating this planning and learning process, R-MAX ends up producing a nearly optimal policy in polynomial time.

---

[3]In Reinforcement Learning, it is common to consider rewards of the form $r(s, a, s')$ that are a function of the action applied, the current state s, and the following state s'. Such rewards, unlike the rewards $r(a, s)$ that we have considered so far, need to be pushed inside the summation as shown in (6.22). When the model parameters are known, as in planning, these 3-place rewards $r(s, a, s')$ can be replaced by the 2-place rewards $r(a, s)$ by setting $r(a, s) = \sum_{s' \in S} P_a(s'|s) r(s, a, s')$.

CHAPTER 7

POMDP Planning: Stochastic Actions and Partial Feedback

Partially observable MDPs (POMDPs) generalize MDPs by allowing states to be partially observable through sensors that map the true state of the world into observable tokens according to known probabilities. A POMDP can be understood as an MDP over belief states where a belief state is a probability distribution over the states. In Goal POMDPs, the task is to reach the goal with certainty given a known initial belief and actions and observations that change the world and the beliefs. In this chapter, we look at Goal and Discounted POMDP models, and at the basic exact and approximate methods for solving them.

7.1 GOAL, SHORTEST-PATH, AND DISCOUNTED POMDPS

The differences between Goal POMDPs and Goal MDPs are in the *initial situation*, that is no longer assumed to be known, and in the *feedback*, that no longer provides full information about the state of the world. Instead, the initial situation is characterized by an initial belief b_0, and observations are characterized by means of a *sensor model* where tokens $o \in O$ are observed with probabilities $P_a(o|s)$ where s is the true but possibly hidden state of world and a is the last action executed. A Goal POMDP is thus given by:

- a finite state space S,

- a probability distribution b_0 over the states such that $b_0(s)$ stands for the probability of s being the true initial state,

- a non-empty subset $S_G \subseteq S$ of *observable* goal states,

- sets of actions $A(s)$ applicable at each state $s \in S$,

- transition probabilities $P_a(s'|s)$ for s' being the next state after action $a \in A(s)$ is applied in state s,

- positive action costs $c(a, s)$ incurred after applying action $a \in A(s)$ in the state s, and

- sensor probabilities $P_a(o|s)$ of receiving observation token $o \in O$ in state s when the last applied action was a.

As for Goal MDPs, rather than regarding goal states as terminal states to be reached, it is often convenient to regard them as absorbing, cost-free states, i.e., states s for which the transition probabilities are $P_a(s'|s) = 1$ iff $s' = s$, and costs are $c(a, s) = 0$ for any action a. In Goal POMDPs, goal states are assumed to be observable so that there is never uncertainty about whether the goal has been reached or not—an assumption that guarantees that any policy that does not reach the goal with certainty incurs in infinite expected cost. We will see that Discounted POMDPs, whether cost or reward-based, can be easily transformed into equivalent Goal POMDPs that obey these restrictions.

The selection of the best action for achieving the goal in a POMDP depends on the observed execution $\langle a_0, o_0, a_1, o_1, \ldots \rangle$. However, in POMDPs, it's no longer the case that the last observation summarizes the previous execution, and that optimal policies map the last observation into an action. This is because, while the dynamics and cost structure of the model are still Markovian as in MDPs, the state of the system is no longer known. Yet, while the last observation does not summarize the past execution, the *probability distribution* over the states does [Astrom, 1965, Smallwood and Sondik, 1973, Striebel, 1965]. This probability distribution is called the *belief state* of the agent. The initial belief is given by the prior probability b_0 in the model, and the beliefs following an execution are defined inductively from it as follows. If b is the belief state before the agent performs an action a, the belief state b_a that results after the action a is:

$$b_a(s) = \sum_{s' \in S} P_a(s|s')b(s') . \tag{7.1}$$

Then if the observation token o is obtained, the belief that results from b after the action a and the observation o, denoted as b_a^o, is:

$$b_a^o(s) = P_a(o|s)b_a(s)/b_a(o) , \tag{7.2}$$

where $b_a(o)$ is the probability of observing o after doing the action a in the belief b:

$$b_a(o) = \sum_{s \in S} P_a(o|s)b_a(s) . \tag{7.3}$$

Equations 7.1 and 7.2 for POMDPs generalize Eqs. 5.1 and 5.2 for the non-deterministic, partial observable models considered in Chapter 5, by taking probabilities into account, both in the system dynamics and in the sensors.

Provided that the set $A(b)$ of actions that are applicable in a belief state b is defined as the set of actions that are applicable in each of the states that are possible according to b, and that the cost $c(a, b)$ of applying an action a in a belief state b is defined as the expected cost

$$c(a, b) = \sum_{s \in S} c(a, s)b(s) , \tag{7.4}$$

it is simple to transform a Goal POMDP M into an equivalent fully observable Goal MDP *over* belief states M' where:

- the states b in M' are the belief states over M,

- the initial state in M' is the belief state b_0,

- the goal states in M' are the *goal beliefs* b_G such that $b_G(s) = 0$ if s is not a goal state in M,

- the set of actions $A(b)$ applicable in b is comprised of the actions a such that $a \in A(s)$ for all s such that $b(s) > 0$,

- the transition probabilities $P_a(b'|b)$ are equal to $b_a(o)$ if $b' = b_a^o$, and otherwise equal to 0, and

- the costs $c(a, b)$ are given by (7.4), and are positive (and bounded away from zero) except when b is a *goal belief* in which case $c(a, b) = 0$.

The solution to this *belief MDP* is a policy mapping belief states into actions that yields a solution to the original POMDP. In particular, the equations determining the expected costs $V^\pi(b)$ of a policy π from belief b are:

$$V^\pi(b) = c(a, b) + \sum_{o \in O} b_a(o) V^\pi(b_a^o) \tag{7.5}$$

for non-goal beliefs b, and $V(b) = 0$ for goal beliefs, while the optimal cost function $V^*(b)$ is the solution of the equation

$$V(b) = \min_{a \in A(b)} [c(a, b) + \sum_{o \in O} b_a(o) V(b_a^o)] \tag{7.6}$$

for non-goal beliefs b, and $V(b) = 0$ for goal beliefs. The problem in solving the belief MDP, however, is that unlike the MDPs considered in the last chapter, it has a continuous and infinite state space given by the probability distributions over the states in the POMDP. The exact methods for solving POMDPs must address this challenge. As for MDPs, we will assume that there are no *dead-end beliefs* b from which goal beliefs cannot be reached, or alternatively, that there is a proper policy π that ensures that a goal belief will be reached from any belief with probability 1.

SHORTEST-PATH AND DISCOUNTED MODELS

As in the case of MDPs, one can define Stochastic Shortest-Path POMDPs as Goal POMDPs where the action costs $c(a, s)$ are not required to be positive over non-goal states s. In such a case, however, for the model and the solutions to be well defined, every policy π that does not achieve the goal with certainty from a belief b must have infinite cost $V^\pi(b)$.

Discounted Cost and Reward-based POMDPs are defined as for MDPs with no requirement on the presence of terminal goal states or in the sign of action costs or rewards. Rather, a discount factor $0 < \gamma < 1$ is used to discount future costs or rewards at a geometric rate. The expected accumulated discounted cost or reward of any policy is then always bounded. As in the case of MDPs (Section 6.1), Discounted Reward POMDPs can be compiled into *equivalent* Goal POMDPs by means of three transformations that also preserve equivalence in the POMDP setting, namely, the addition of a constant reward R to make all rewards negative, the transformation of rewards to be maximized into costs to be minimized, and the elimination of the discount factor by the addition of a dummy, observable goal state [Bonet and Geffner, 2009].

7.2   EXACT OFFLINE ALGORITHMS

Since the number of belief states in POMDPs is infinite, policies and value functions cannot be stored explicitly in memory. This is the first obstacle when trying to solve a POMDP with methods like Value or Policy Iteration. Indeed, in Value Iteration, the full parallel DP update should map a value vector V_k over all beliefs b into the value vector V_{k+1}:

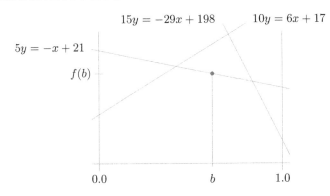

Figure 7.1: Example of a piecewise linear and concave (*pwlc*) function $f(b)$ over beliefs b. The function f is represented by a finite set Γ of $|S|$-dimensional vectors such that $f(b) = \min_{\alpha \in \Gamma} \sum_{s \in S} b(s)\alpha(s)$. In the example, S contains two states and Γ contains three planes (lines). Every belief state b on S corresponds to a point in the interval $[0, 1]$ and the value $f(b)$ is determined by the projection of the point b on the concave hull of Γ.

$$V_{k+1}(b) = \min_{a \in A(b)} c(a, b) + \sum_{o \in O} b_a(o) V_k(b_a^o) . \tag{7.7}$$

Yet such an update cannot be implemented by iterating over all possible belief states. In 1973, however, Sondik observed that if Value Iteration starts from a *piecewise linear and concave (pwlc) function* V_0, then all functions V_k resulting from these updates remain *pwlc*. A *pwlc* function f on the continuous belief space over S is a combination of linear functions given as

$$f(b) = \min_{\alpha \in \Gamma} \sum_{s \in S} b(s)\alpha(s) \tag{7.8}$$

where Γ is a finite set of $|S|$-dimensional real vectors. The *pwlc* function f can be stored in finite space as the set of vectors Γ. Figure 7.1 shows an example of a *pwlc* function. Clearly, any constant function $f(b) \equiv v$ is a *pwlc* function that corresponds to the singleton $\Gamma = \{\alpha\}$ where the vector α is such that $\alpha(s) = v$ for all states s.

Sondik's result is fundamental as it provides a feasible way for implementing Value Iteration: starting from a set Γ_0 of vectors that represent the *pwlc* value function V_0, a new set of vectors Γ_1 is computed that represents the *pwlc* value function V_1 that results from a full DP update of V_0, and so on, until a value function V_k is obtained with residuals that do not exceed a given ϵ. We will follow Sondik in showing how these updates can be carried out. We use a formulation with a *discount factor* γ, $0 < \gamma < 1$, that ensures convergence to a given residual in a bounded number of iterations, and assume that all actions $a \in A$ are *applicable* in all states so that $A(b) = A$ for any belief b. We also assume a *cost setting*; for rewards, costs $c(a, s)$ must be replaced by rewards $r(a, s)$, and minimizations by maximizations.

Given an initial *pwlc* value function V_0, and assuming inductively that the value function V_k can be characterized by a set of vectors Γ_k as:

$$V_k(b) = \min_{\alpha \in \Gamma_k} \sum_{s \in S} b(s)\alpha(s) , \tag{7.9}$$

we need to show that the function V_{k+1}, given as the full DP update of the function V_k:

$$V_{k+1}(b) = \min_{a \in A} [c(a,b) + \gamma \sum_{o \in O} b_a(o) V_k(b_a^o)] \tag{7.10}$$
$$= \min_{a \in A} [c(a,b) + \gamma \sum_{o \in O} b_a(o)[\min_{\alpha \in \Gamma_k} \sum_{s \in S} b_a^o(s)\alpha(s)]] . \tag{7.11}$$

is also a *pwlc* function, given by a finite set of vectors Γ_{k+1} defined in terms of the vectors in Γ_k and the parameters of the POMDP.

Notice that for each observation o, the outer sum in the right-hand side of (7.11) "picks" a vector α for the inner sum, and that such vectors can be summarized with a *choice function* $v : O \to \Gamma_k$. Moreover, since the outer sum picks vectors that minimize the inner sum, the min inside can be pulled out by converting it into a minimization over the collection $V_k \stackrel{\text{def}}{=} \{v \,|\, v : O \to \Gamma_k\}$ of all the $|\Gamma_k|^O$ choice functions. If v is one such choice function and $v(o) = \alpha$, the notation $v(o)(s)$ below stands for $\alpha(s)$:

$$V_{k+1}(b) = \min_{a \in A} [c(a,b) + \min_{v \in V_k} \gamma \sum_{o \in O} b_a(o) \sum_{s \in S} b_a^o(s)v(o)(s)] \tag{7.12}$$
$$= \min_{a \in A} [c(a,b) + \min_{v \in V_k} \gamma \sum_{s \in S} \sum_{o \in O} v(o)(s)b_a(o)b_a^o(s)] . \tag{7.13}$$

Making use of the definitions of the probabilities $b_a^o(s)$ and $b_a(o)$ in Eqs. 7.2 and 7.3 respectively, it follows that:

$$V_{k+1}(b) = \min_{a \in A} [c(a,b) + \min_{v \in V_k} \gamma \sum_{s,o} v(o)(s) \sum_{s'} b(s') P_a(s|s') P_a(o|s)] \tag{7.14}$$
$$= \min_{a \in A} \min_{v \in V_k} [c(a,b) + \gamma \sum_{s,o} v(o)(s) \sum_{s'} b(s') P_a(s|s') P_a(o|s)] \tag{7.15}$$

and by plugging the definition of $c(a,b)$ and regrouping terms, that

$$V_{k+1}(b) = \min_{a \in A, v \in V_k} \sum_{s'} b(s')[c(a,s') + \gamma \sum_{s,o} v(o)(s) P_a(s|s') P_a(o|s)] \tag{7.16}$$
$$= \min_{\alpha \in \Gamma_{k+1}} \sum_{s \in S} b(s)\alpha(s) \tag{7.17}$$

where $\Gamma_{k+1} \stackrel{\text{def}}{=} \{\alpha_{a,v} \,|\, a \in A, v \in V_k\}$ is the collection of vectors $\alpha_{a,v}$ defined by

$$\alpha_{a,v}(s) \stackrel{\text{def}}{=} c(a,s) + \gamma \sum_{s',o} v(o)(s') P_a(s'|s) P_a(o|s') . \tag{7.18}$$

The set Γ_{k+1} contains at most $|A|\,|\Gamma_k|^{|O|}$ different vectors $\alpha_{a,v}$, one for each action and choice function $v : O \to \Gamma_k$. However, some of these vectors are dominated by others in the sense that they do not yield the minimum at any belief b. Such vectors can be identified by solving a linear program and removed [Kaelbling et al., 1998]. Similarly, a linear program can be used to check if the residual of the value function represented by Γ_k falls below ϵ for terminating Value Iteration. Figure 7.2 shows the result of applying a full DP update over the function shown in Figure 7.1.

In spite of the exponential complexity of the full DP update, Sondik's representation is ubiquitous in exact and approximated methods, including recent state-of-the-art algorithms that use it in a more selective type of updates, known as *point-based updates*.

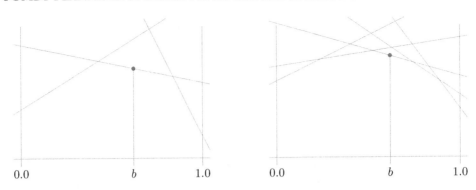

Figure 7.2: The left panel shows the concave hull for a set Γ of vectors that define a *pwlc* value function (cf. Figure 7.1). The right panel shows the result of applying a full update on Γ as described in the text. Only the non-dominated vectors on both sets are shown.

7.3 APPROXIMATE AND ONLINE ALGORITHMS

The complexity of solving POMDPs has limited the applicability of exact algorithms to large problems where approximation methods are used instead. We review some of these methods below.

POINT-BASED BACKUP ALGORITHMS

The exponential blow up in the number of vectors that results from a single DP update is a consequence of updating the value function over all beliefs. An alternative is to update the value function at a restricted subset of belief points, generating fewer vectors, and hence keeping the size of the value function representation smaller. Some of the state-of-the-art algorithms for POMDPs are based on this idea of *point-based value updates*.

If V is a *pwlc* function given by a set of vectors Γ, a *point-based update* or *backup* of V over a set of belief points F refers to the *pwlc* value function $\widehat{V}$ given by a new set of vectors $\widehat{\Gamma}$ such that $\widehat{V}(b) = V_{\text{fb}}(b)$ for every $b \in F$, where V_{fb} is the full backup on V. If F is the whole belief space, $\widehat{V}$ is equal to V_{fb}; otherwise the size of F can be used to control the complexity of the point-based backup. Point-based POMDP algorithms [Pineau et al., 2006, Shani et al., 2012] compute the new set of vectors $\widehat{\Gamma}$ from Γ by adding one vector $backup(V, b)$ for each belief b in F. The vector $backup(V, b)$ is the one that assigns the value $V_{\text{fb}}(b)$ to b:

$$backup(V, b) = \operatorname{argmin}_{\alpha \in \Gamma_{\text{fb}}} \sum_{s \in S} b(s)\alpha(s) \tag{7.19}$$

where $\Gamma_{\text{fb}} = \{\alpha_{a,v} \mid a \in A, v \in \mathcal{V}\}$ is the set of vectors for V_{fb} and $\mathcal{V} = \{v \mid v : O \to \Gamma\}$ is the set of choice functions for Γ. This expression however requires the computation of the set Γ_{fb} whose size $O(|A| |\Gamma|^{|O|})$ is exponential in the number of possible observations. The method below computes the single vector $backup(V, b)$ for an arbitrary belief b in polynomial time. For this, observe first that the sum on the right-hand side of the Bellman equation for updating $V(b)$ can be expressed as:

$$\sum_{o \in O} b_a(o) V(b_a^o) = \sum_{o \in O} b_a(o)\left[\min_{\alpha \in \Gamma} \sum_{s \in S} b_a^o(s)\alpha(s)\right] \tag{7.20}$$

$$= \sum_{o \in O} \left[\min_{\alpha \in \Gamma} \sum_{s,s',s'' \in S} P_a(o|s'') P_a(s''|s') b(s') \alpha(s) \right] \tag{7.21}$$

$$= \sum_{o \in O} \left[\min_{\alpha \in \Gamma} \sum_{s' \in S} g_{a,o}^\alpha(s') b(s') \right] \tag{7.22}$$

where $g_{a,o}^\alpha$ is the vector defined as

$$g_{a,o}^\alpha(s) \overset{\text{def}}{=} \sum_{s' \in S} \alpha(s') P_a(o|s') P_a(s'|s). \tag{7.23}$$

If we write the inner sum as a dot product and observe that

$$\min_{\alpha \in \Gamma} g_{a,o}^\alpha \cdot b = [\operatorname{argmin}\{g_{a,o}^\alpha \cdot b \mid g_{a,o}^\alpha, \alpha \in \Gamma\}] \cdot b, \tag{7.24}$$

we obtain

$$\sum_{o \in O} b_a(o) V(b_a^o) = \sum_{o \in O} \left[\min_{\alpha \in \Gamma} g_{a,o}^\alpha \cdot b \right] \tag{7.25}$$

$$= \sum_{o \in O} \left\{ [\operatorname{argmin}\{g_{a,o}^\alpha \cdot b \mid g_{a,o}^\alpha, \alpha \in \Gamma\}] \cdot b \right\} \tag{7.26}$$

$$= \left[\sum_{o \in O} \operatorname{argmin}\{g_{a,o}^\alpha \cdot b \mid g_{a,o}^\alpha, \alpha \in \Gamma\} \right] \cdot b \tag{7.27}$$

where the sum in (7.27) is a (point-wise) sum of vectors. Finally, if $g_{a,b}$ is defined as the vector with entries

$$g_{a,b}(s) \overset{\text{def}}{=} c(a,s) + \gamma \sum_{o \in O} g_{a,o}^b(s) \tag{7.28}$$

where the vector $g_{a,o}^b$ is the one that minimizes the scalar product $g_{a,o}^\alpha \cdot b$ for $\alpha \in \Gamma$. Then, *the value of the full backup of V at the belief b* becomes:

$$V_{\text{fb}}(b) = \min_{a \in A} c(a,b) + \gamma \left[\sum_{o \in O} \operatorname{argmin}\{g_{a,o}^\alpha \cdot b \mid g_{a,o}^\alpha, \alpha \in \Gamma\} \right] \cdot b \tag{7.29}$$

$$= \min_{a \in A} \left[g_{a,b} \cdot b \right] \tag{7.30}$$

so that

$$backup(V,b) = \operatorname{argmin}_{g_{a,b}, \ a \in A} \left[g_{a,b} \cdot b \right]. \tag{7.31}$$

This derivation provides an efficient method for computing the vector that encodes the update of the value function V at the belief point b. There are indeed $|A|$ vectors $g_{a,b}$, each one of which can be computed in $O(|S||O||\Gamma|)$ time provided that the vectors $g_{a,o}^\alpha$ are precomputed and stored, an operation that requires $O(|A||O||S|)$ time. Moreover, the vectors $g_{a,o}^\alpha$ do not depend on the belief point b and thus, once computed, can be reused when computing the backup over other beliefs.

The different *point-based POMDP algorithms* differ mainly on the set of beliefs F selected for update in each iteration, in the initial set of vectors, and in the termination condition [Shani et al., 2012]. Like the RTDP algorithm for POMDPs below, recent point-based algorithms aim to exploit the information about the initial belief state, use admissible value functions, and focus on the belief states that are reachable from the initial belief state following greedy policies.

RTDP-Bel

% *Initial value function V given by heuristic h*
% *Changes to V stored in a hash table using discretization function $d(\cdot)$*

Let $b := b_0$ the initial belief
Sample state s with probability $b(s)$
While b is not a goal belief **do**
    Evaluate each action $a \in A(b)$ as: $Q(a, b) := c(a, b) + \sum_{o \in O} b_a(o) V(b_a^o)$
    Select best action $\mathbf{a} := \operatorname{argmin}_{a \in A(b)} Q(a, b)$
    Update value $V(b) := Q(\mathbf{a}, b)$
    Sample next state s' with probability $P_{\mathbf{a}}(s'|s)$ and set $s := s'$
    Sample observation o with probability $P_{\mathbf{a}}(o|s)$
    Update current belief $b := b_a^o$
end while

Figure 7.3: RTDP-Bel is RTDP over the belief MDP with an additional provision: for reading or writing the value $V(b)$ in the hash table, b is replaced by $d(b)$ where d is a discretization function.

RTDP-BEL

RTDP-Bel [Bonet and Geffner, 2009, Geffner and Bonet, 1998] is a direct adaptation to Goal POMDPs of the RTDP algorithm developed for Goal MDPs [Barto et al., 1995] reviewed in Chapter 6, where states are replaced by belief states and the updates are done using the expression (7.7) for POMDPs. The code for RTDP-Bel is shown in Figure 7.3. There is just one difference between RTDP-Bel and RTDP: in order to bound the size of the hash table and make the updates more effective, each time that the hash table is accessed for reading or writing the value $V(b)$, the belief b is *discretized*. The discretization function d maps each entry $b(s)$ into the entry $d(b(s)) = \operatorname{ceil}(D \cdot b(s))$ where D is a positive integer (the discretization parameter), and $\operatorname{ceil}(x)$ is the least integer greater than or equal to x. For example, if $D = 10$ and b is the vector $(0.22, 0.44, 0.34)$ over the states $s \in S$, $d(b)$ is the vector $(3, 5, 4)$. The discretization is used in the operations for accessing the hash table and does not affect the beliefs that are generated during a trial. Using a terminology that is common in Reinforcement Learning, the discretization is a *function approximation* device [Bertsekas and Tsitsiklis, 1996, Sutton and Barto, 1998], where a single parameter, the value stored at cell $d(b)$ in the hash table, is used to represent the value of all beliefs b' that discretize into $d(b') = d(b)$. This approximation relies on the assumption that the value of beliefs that are close, should be close as well. Moreover, the discretization preserves supports (the states s with $b(s) > 0$) and never collapses the values of two beliefs if there is a state that is excluded by one but not by the other.

Belief discretization makes the value function representation finite at the cost of theoretical properties that do not carry automatically from RTDP to RTDP-Bel. First, convergence of RTDP-Bel is not guaranteed and actually the value in a cell may oscillate. Second, the value function approximated in this way does not remain necessarily a lower bound.

RTDP and RTDP-Bel can be used both for offline and online planning. For Goal MDPs, RTDP trials are guaranteed to reach a goal state provided that there are no dead-ends in the problem. The same is usually true for RTDP-Bel but this cannot be guaranteed due the approximation introduced by the discretization. In any case, if the input problem is a Discounted POMDP, whether reward or cost-based, it must first be converted into an equivalent Goal POMDP before RTDP-Bel is run. This transformation has been applied to the existing benchmarks for Discounted Reward-based POMDPs in order to compare RTDP-Bel with point-based POMDP algorithms [Bonet and Geffner, 2009].

PO-UCT

PO-UCT is in turn a generalization of the UCT algorithm for POMDPs [Silver and Veness, 2010]. Adapting UCT to POMDPs is less direct than adapting RTDP because, as a model-free algorithm, PO-UCT cannot keep track of the exact beliefs. PO-UCT thus keeps track of *executions* or *histories*, sequences of actions and observations $h = \langle a_0, o_0, a_1, o_1, \ldots, a_k, o_k \rangle$, and for each history h, it approximates the belief that would result from such an execution and a given initial belief state b_0, by a set of state samples $B(h)$. The nodes in the tree built by PO-UCT refer to executions h' that extend the real execution h, starting with the empty execution $h_0 = \langle \rangle$. If h is the real execution so far, PO-UCT performs a number of simulations starting in states s sampled from $B(h)$, applies the action a that minimizes the costs $V(ha)$, gets the observation o, and resumes the loop with the history hao. As in UCT, the nodes h are associated with two fields in addition to the set of samples $B(h)$: the value $V(h)$ associated to the execution, and a counter $N(h)$ that tracks the number of simulations that have passed through the node h.

The planning tree is expanded by performing simulations that start at states s sampled from the belief $B(h)$ associated with the real execution h so far. Actions are selected in a node h of the tree as in UCT, using the current values $V(ha)$ of the actions a in h, and the number of visits $N(h)$ and $N(ha)$. When the action a is performed in a state s, the simulator returns an observation o, a next state s', and a sampled cost c. If the resulting node hao is not in the tree, it is added to the tree, and a rollout of the base policy is used to initialize the value of the node and to update the ancestor nodes using Monte-Carlo updates. If the node hao is in the tree, the same process is applied from that node and the associated state s'. In either case, the counter $N(hao)$ is adjusted and the sample s' is added to $B(hao)$.

When the planning episode finishes and an action a is selected for execution following the current execution h, the action is executed and an observation o is gathered. The next planning episode starts from the resulting history hao. Nodes h' in the tree that are not extensions of the execution so far can be pruned. Code for a cost-based version of the PO-UCT algorithm is shown in Figure 7.4. A problem for PO-UCT arises from the way it approximates beliefs by samples, which does not prevent reaching real executions h with very few samples $B(h)$ that limit the information that can be obtained by planning from h. By using domain-specific methods for adding new samples in such cases, PO-UCT has been shown to exhibit excellent performance over a collection of large POMDPs, including games such as Battleship and a partially observable version of Pacman [Silver and Veness, 2010].

7.4   BELIEF TRACKING IN POMDPS

Like in the logical partially observable setting (Section 5.5), belief tracking in POMDPs is *intractable* for problems represented in compact form, with Eqs. 7.1 and 7.2 defining a plain belief tracking algorithm for computing the next belief b_a^o from the belief b, the action a, and the observation o, that

---

SEARCH(h)

  repeat

      Sample s according to b_0 if $h = \langle \rangle$ or $B(h)$ otherwise

      SIMULATE($s, h, 0$)

  until time is up

  return $\operatorname{argmin}_a V(\langle ha \rangle)$

SIMULATE($s, h, depth$)

  if $\gamma^{depth} < \epsilon$ **then return** 0

  if h does not appear in tree T **then**

      for all action $a \in A$ **do**

         Insert $\langle ha \rangle$ in tree as $T(\langle ha \rangle) := \langle 0, 0, \emptyset \rangle$

      end for

      return ROLLOUT($s, h, depth$)

  end if

  $a^* := \operatorname{argmin}_a V(\langle ha \rangle) - C \sqrt{\log N(h) / N(\langle ha \rangle)}$

  Sample (s', o, c) using simulator with state s and action a^*

  $Cost := c + \gamma \cdot$ SIMULATE($s', \langle hao \rangle, 1 + depth$)

  $B(h) := B(h) \cup \{s\}$

  Increment $N(h)$ and $N(\langle ha \rangle)$

  $V(\langle ha \rangle) := V(\langle ha \rangle) + [Cost - V(\langle ha \rangle)] / N(\langle ha \rangle)$

  return $Cost$

ROLLOUT($s, h, depth$)

  if $\gamma^{depth} < \epsilon$ **then return** 0

  Let $a := \pi(h)$

  Sample (s', o, c) using simulator with state s and action a

  return $c + \gamma \cdot$ ROLLOUT($s', \langle hao \rangle, 1 + depth$)

Figure 7.4: PO-UCT algorithm for Discounted Cost-based POMDPs. Each node in the tree corresponds to an execution history h that is associated with a triplet $\langle N(h), V(h), B(h) \rangle$ made up of a counter $N(h)$, a value $V(h)$ for the node, and a set of samples $B(h)$. The policy π is the base policy used in PO-UCT. The action selected for execution after the history h is the one returned by SEARCH(h). This action a is applied, the observation o is obtained, and the process resumes with $h := hao$.

is exponential in the number of problem variables. The problem of computing the belief that results at time $k + 1$ from a given execution $h = \langle a_0, o_0, \ldots, a_k, o_k \rangle$, starting in a given initial belief b_0, can be expressed as a probabilistic inference problem over a Dynamic Bayesian Network [Pearl, 1988, Russell and Norvig, 2009]. Exact probabilistic inference over Bayesian Networks is exponential in a parameter associated with the underlying directed graph, known as the *treewidth*, which is related to the maximum number of variables in the network that have to be collapsed into a single variable so that the result is a Bayesian Tree. Since often the treewidth is not bounded, approximation algorithms are common. In the case of Dynamic Bayesian Networks (DBN), a usual algorithm is *particle filtering* where beliefs are approximated by a set of states or particles [Doucet et al., 2000]. In its most basic form, given a set of samples B_k providing an approximate representation of the belief b_k after an execution $h_k = \langle a_0, o_0, \ldots, a_k, o_k \rangle$, a new set of samples B_{k+1} can be obtained for approximating the belief b_{k+1} that results from the action a_{k+1} and the observation o_{k+1}, by the following three steps. First, each sample state s_k in B_k is propagated into a state s_{k+1} sampled with the transition probability $P_{a_k}(s_{k+1}|s_k)$. Second, the new samples s_{k+1} are assigned a weight given by the observation probability $P_{a_k}(o_{k+1}|s_{k+1})$. Third, the set of weighted samples is resampled to yield the set of unweighted samples B_{k+1} [Russell and Norvig, 2009]. The initial set of samples B_0 is obtained by sampling the initial belief b_0. The probability that a given formula is true at time $k + 1$ is obtained from the ratio of samples in B_{k+1} where the formula is true. Particle filtering does best when there are few zero entries in the transition and observation probabilities. The PO-UCT algorithm above approximates beliefs from histories using a particle filter of this type.

7.5   OTHER MDP AND POMDP SOLUTION METHODS

We close this chapter by pointing to other solution methods for MDPs and POMDPs.

Finite-State Controllers. When the value function is expressed by a set of vectors, Value Iteration results in policies that can be understood as finite-state controllers. One of the advantages of such policies as opposed to policies represented as functions over beliefs, is that the former do not require keeping track of beliefs. Following Kaelbling et al. [1998], one can compute an automaton M_k for solving the k-horizon version of a Discounted POMDP. Such an automaton consists of controller states associated with actions a, and transitions that map observations o into other controller states. The automata $M_0, M_1, \ldots, M_{k+1}$ are constructed iteratively in a way that M_{k+1} is obtained from the set Γ_{k+1} of vectors defining the value function V_{k+1} and the previous automaton M_k. Recall that each vector in Γ_{k+1} is associated with at least one action a and a choice function v in $\mathcal{V}_k$ through Eq. 7.18. Thus, a vector $\alpha_{a,v}$ in Γ_{k+1}, where $v : O \to \Gamma_k$, can be understood as prescribing the action a and picking up a vector in Γ_k for each observation o. Therefore, starting from a singleton Γ_0 and automaton M_0, the automaton M_{k+1} is built by adding $|\Gamma_{k+1}|$ new nodes to M_k, one for each vector $\alpha_{a,v} \in \Gamma_{k+1}$, setting as a the action to do at nodes corresponding to vectors $\alpha_{a,v}$, and adding transitions to the nodes $v(o)$ in M_k for each observation o. In any of these automata M_{k+1}, the initial execution node for a given belief b is the one associated with the vector $\mathrm{argmin}_{\alpha \in \Gamma_k} \alpha \cdot b$. If k is large enough so that the residual of V_k is less than ϵ, then the automaton M_k defines a *non-stationary* policy that is ϵ-optimal. While this approach for computing controllers for POMDPs has the same computational limitations as Value Iteration, it suggests other approaches that can be understood as a form of Policy Iteration implemented on a class of controllers [Hansen, 1998, Poupart and Boutilier, 2003]. More recent approaches cast the problem of devising the best finite-state controller with a given number N

of controller states, as a non-linear optimization problem [Amato et al., 2010]. Solutions of this form are common in multiagent POMDP models known as Decentralized POMDPs or DEC-POMDPs [Bernstein et al., 2002].

Symbolic Methods for MDPs and POMDPs. A value function over states that assigns values x to variables X can be often expressed in compact form as a value function over formulas built from the atoms $X = x$. The set of entries $V(s) = v$ that share the same value v can then be expressed as a single entry $V(\varphi) = v$ where φ is the formula that is true precisely in the states s such that $V(s) = v$. This is the basic idea exploited in symbolic approaches that make use of data structures known as algebraic decision diagrams for representing and operating on value function of this type. Symbolic methods have been developed for problems with a finite number of states [Hoey et al., 1999] and for problems with an infinite number of states [Boutilier et al., 2001, Sanner and Boutilier, 2009]. The common principle is the compact representation of value functions, a feature that is often independent of the planning algorithm. In addition, MDP and POMDP algorithms have been transformed into symbolic algorithms by expressing their basic operations in terms of the symbolic representation [Feng and Hansen, 1999, Feng et al., 2002, Sanner and Kersting, 2010]. Symbolic methods have also been used in model-checking and non-probabilistic planning [Cimatti et al., 2003, Clarke et al., 2000, Edelkamp and Schrödl, 2012].

Finite-horizon POMDPs and Probabilistic Inference. In analogy to the SAT approach to deterministic planning, it is possible to map POMDP problems over finite horizons into probabilistic inference problems over Dynamic Bayesian Networks [Attias, 2003, Botvinick and An, 2008, Toussaint and Storkey, 2006]. While the reduction of planning to inference, deductive or probabilistic, is conceptually appealing, the scalability-quality tradeoff in the probabilistic case, unlike the SAT approach in the deductive case, is yet to be analyzed.

CHAPTER 8

Discussion

The selection of the action to do next is one of the central problems faced by autonomous agents. As discussed in Chapter 1, the problem is normally addressed in three different ways: in the *hardwired approach*, the control is set by nature or by a programmer, in the *learning-based approach*, the control is learned by trial-and-error, in the *model-based approach*, the control is derived from a model of the actions, sensors, and goals. Planning is the model-based approach to autonomous behavior, and in this book we have considered the main planning models and methods. In this last chapter, we list some challenges in current planning research, and discuss briefly how the work in scalable computational models of planning can contribute to the understanding of one of the most unique human features, namely, the ability to plan, often in the context of other agents that have goals and make plans too.

8.1 CHALLENGES AND OPEN PROBLEMS

There are many open problems in planning research. We list a few that we regard as particularly relevant and important.

Multiagent Planning. One important open problem is planning in the presence of other agents that plan, often called multiagent planning. People do this naturally all the time: walking on the street, driving, etc. The first question is how plans should be defined in that setting. This is a subtle problem and many proposals have been put forward, often building on equilibria notions from game theory [Bowling et al., 2003, Brafman et al., 2009], yet currently there are no models, algorithms, and implementations of domain-independent planners able to plan meaningfully and efficiently in such settings. This is probably not too surprising given the known limitations of game theory as a descriptive theory of human behavior. It is possible indeed that domain-independent multiagent planners based on a narrow view of human rationality that ignores the social dimension, cannot get off the ground. Planners able to capture the interactions that appear in simple tales such as Little Red Riding Hood may have to be built on different foundations. There has been work in AI on models for multiagent planning like DEC-POMDPs [Bernstein et al., 2002], but the models are too complex computationally, and yet lack crucial features such as the need for representing the beliefs of other agents. The limitations suggest the need to explore other formulations as well.

Factored and Scalable POMDP Planning. Terms like "factored MDPs" and "factored POMDPs" have been used to refer to MDP and POMDP models represented in compact form over a set of variables. Often, however, general MDP and POMDP algorithms, whether based on dynamic programming or heuristic search, throw the variables away and just deal with states or belief states, the exception being symbolic methods. The situation is different in classical planning and in most work in contingent and conformant planning that either leverage on classical planners or use heuristics extracted from the compact representation to guide the search to the goal. In this sense, there is discontinuity in the book, with Chapters 2–5 building on top of each other, and Chapters 6–7 on MDPs and POMDPs starting almost from scratch. Yet, the only difference between contingent and POMDP models is that the former represent uncertainty by means of sets, and the latter by means of probability distributions. Probabilistic methods, both model-based and model-free, have often appealed to function approximation schemes for representing the value function and for scaling up, yet such methods do not compete, for example, with classical planners in deterministic problems. There is thus a computational gap to be bridged between logical and probabilistic methods. Planners such as FF-Replan have been shown to be quite effective for a wide range of MDPs by ignoring the actual probability values and using relaxations into classical planning for selecting the action to do next. Of course, this is a rough way to exploit compact representations computationally, yet it cannot be neglected. In particular, approaches of this type, as well as similar approaches developed for partially observable problems, can be used to generate suboptimal but meaningful base policies for MDPs and POMDPs that can then be improved in anytime optimal fashion by UCT-like algorithms. Other options are certainly possible.

Learning to search. Learning can play several roles in model-based approaches, the first of which is learning the model itself from experience and partial observations (see below). Learning, however, has also a role to play in the search for solutions; a role that has been crucial in the context of SAT [Biere et al., 2012], but has not been fully exploited in planning except in the context of planning as SAT [Kautz and Selman, 1996]. For example, consider an agent that has to deliver a large package to one of two cells A or B in a grid, by going to the cell and dropping the package. Furthermore, assume that A is closer to the agent than B but A cannot be entered while holding a large package. Most current classical planning heuristics will drive the search toward A in a way resembling a fly that wants to get past a closed window. Unlike flies, however, search algorithms avoid revisiting the same states, and would eventually solve the problem after partially exhausting the space around A. A more intelligent strategy would be to note that the failed search around A is the result of an interaction ignored by the heuristic that should be fixed. This is precisely what SAT solvers do: they identify the causes for failure and fix them while searching. Traditional heuristic methods cannot replicate this behavior because they ignore the structure of the heuristic function, yet this structure is available to heuristic search methods in planning that should be able to exploit it. This same limitation applies to heuristic search algorithms like LRTA* and those used for solving MDPs and POMDPs: the values of states and belief states are learned very slowly because there is no analysis for explaining what was wrong with the updated estimates. It's an open question whether something akin to the *conflict-directed learning* from SAT could be used in a cost-effective way in the setting of heuristic search.

Generalized Planning. The problem of model and feature learning are related to the problem of generalized planning reviewed in Section 1.5 where a policy is sought not just for one planning instance but for many instances, e.g., all block world instances. Often general policies of this type can be expressed in a compact way provided the right features. The question is how to get simultaneously the right features and the policies. One approach that has been pursued to do this learns compact policies from examples using features obtained from the potentially infinite collection of predicates defined by a domain-independent grammar and a given set of primitive domain predicates [Fern et al., 2003]. This is an inductive, learning-based approach. An open question is whether these types of compact, generalized policies can also be synthesized from a model by suitable transformations of the problem. The derivation of finite-state controllers using planners considered in Section 4.4 goes in this direction. Also, for example, the generalized planning problem of picking up a green block from a tower of blocks of any size can be cast as a non-deterministic partially observable planning problem over integer variables, that can be modeled and solved with the methods developed by Srivastava et al. [2011b].

Hierarchies. Hierarchies form a basic component of Hierarchical Task Networks (HTNs), an alternative model for planning that is concerned with the encoding of strategies for solving problems (Section 3.11). Hierarchies, however, play no role in state-of-the-art domain-independent planners that are completely flat. Yet, it is clear that most real plans involve primitive actions that can be executed along with high level actions that are abstractions of those. For example the action of picking up a block involves displacements of the robot gripper that must be opened and closed on the right block. A basic question that has not been fully answered yet is how these abstractions can be formed automatically, and how they are to be used to speed up the planning process. For instance, the standard blocks world is an abstraction of a problem where blocks are at certain locations, and the gripper has to move between locations. This abstraction, however, is not adequate when the table has no space for all the blocks, or when the gripper cannot get past towers of a certain height. The open question is how to automatically compile detailed planning descriptions into more abstract ones that can be used to solve the problems more effectively. There is a large body of work on abstract problem solving that is relevant to this question [Bacchus and Yang, 1994, Jonsson, 2007, Knoblock, 1990, Korf, 1987, Marthi et al., 2007, McIlraith and Fadel, 2002, Sacerdoti, 1974], but none so far that solves this problem in a general manner.

Model Learning. We have discussed briefly model-based reinforcement learning algorithms that actively learn model parameters such as probabilities and rewards, yet a harder problem is learning the states themselves from partial observations. Several attempts to generalize reinforcement learning algorithms to such setting have been made, some of which learn to identify useful features and feature histories [Veness et al., 2011], but none so far that can come up with the states and models themselves in a robust and scalable manner from streams of observations and actions.

8.2   PLANNING, SCALABILITY, AND COGNITION

The relevance of the early work in artificial intelligence to cognitive science was based on *intuition*: programs provided a way for specifying intuitions precisely and for trying them out. The more recent work on *domain-independent solvers* in AI is more technical and experimental, and is focused not as much on reproducing intuitions but on *scalability*. This may give the impression that recent work in AI is less relevant to cognitive science than work in the past. This impression, however, may prove to be wrong for two reasons. First, intuition is not what it used to be, and it is now regarded as the tip of an iceberg whose bulk is made of massive amounts of shallow, fast, but unconscious inference mechanisms that cannot be rendered explicit [Gigerenzer, 2007, Hassin et al., 2005, Kahneman, 2011, Wilson, 2002]. Second, whatever these mechanisms are, they appear to work pretty well and to scale up. This is no small feat, given that most methods, whether intuitive or not, do not. By focusing on the study of meaningful models and the computational methods for dealing with them *effectively*, AI may prove its relevance to the understanding of human cognition in ways that may go well beyond the rules, cognitive architectures, and knowledge structures of the 80s. Human cognition, indeed, still provides the inspiration and motivation for a lot of research in AI. The use of Bayesian Networks in developmental psychology for understanding how children acquire and use causal relations [Gopnik et al., 2004], and the use of reinforcement learning algorithms in neuroscience for interpreting the activity of dopamine cells in the brain [Schultz et al., 1997], are two examples of general AI techniques that have made it recently into cognitive science. As AI focuses on models and solvers able to scale up, more techniques are likely to follow. In this short book, we have reviewed work in computational models of planning with an emphasis on deterministic planning models where automatically derived relaxations and heuristics manage to integrate information about the current situation, the goal, and the actions for directing an agent effectively toward a goal. The computational model of goal appraisals that is based on the solution of low-polynomial relaxations may shed light on the computation, nature, and role of other type of appraisals, and on why appraisals are opaque to cognition and cannot be rendered conscious or articulated in words.[1]

---

[1]This section is taken from Geffner [2010, 2013b] where these issues are discussed in more detail.

Bibliography

P. E. Agre and D. Chapman. Pengi: An implementation of a theory of activity. In *Proc. 6th Nat. Conf. on Artificial Intelligence*, pages 268–272, 1987. 12

A. Albarghouthi, J. A. Baier, and S. A. McIlraith. On the use of planning technology for verification. In *Proc. ICAPS'09 Workshop VV&PS*, 2009. 64

A. Albore, H. Palacios, and H. Geffner. A translation-based approach to contingent planning. In *Proc. 21st Int. Joint Conf. on Artificial Intelligence*, pages 1623–1628, 2009. 57, 72, 73, 74

A. Albore, M. Ramírez, and H. Geffner. Effective heuristics and belief tracking for planning with incomplete information. In *Proc. 21st Int. Conf. on Automated Planning and Scheduling*, pages 2–9, 2011. 72

C. Amato, D. S. Bernstein, and S. Zilberstein. Optimizing fixed-size stochastic controllers for POMDPs and decentralized POMDPs. *Journal of Autonomous Agents and Multi-Agent Systems*, 21(3):293–320, 2010. DOI: 10.1007/s10458-009-9103-z 108

E. Amir and B. Engelhardt. Factored planning. In *Proc. 18th Int. Joint Conf. on Artificial Intelligence*, 2003. 35

K. Astrom. Optimal control of Markov Decision Processes with incomplete state estimation. *Journal of Mathematical Analysis and Applications*, 10:174–205, 1965. 98

H. Attias. Planning by probabilistic inference. In *Proc. 9th Int. Workshop on Artificial Intelligence and Statistics*, 2003. 108

P. Auer, N. Cesa-Bianchi, and P. Fischer. Finite-time analysis of the multiarmed bandit problem. *Machine Learning*, 47(2):235–256, 2002. DOI: 10.1023/A:1013689704352 94

F. Bacchus and F. Kabanza. Using temporal logics to express search control knowledge for planning. *Artificial Intelligence*, 116:123–191, 2000. DOI: 10.1016/S0004-3702(99)00071-5 63

F. Bacchus and Q. Yang. Downward refinement and the efficiency of hierarchical problem solving. *Artificial Intelligence*, 71:43–100, 1994. DOI: 10.1016/0004-3702(94)90062-0 111

C. Bäckström and B. Nebel. Complexity results for SAS+ planning. *Computational Intelligence*, 11(4):625–655, 1995. DOI: 10.1111/j.1467-8640.1995.tb00052.x 25

J. A. Baier, F. Bacchus, and S. A. McIlraith. A heuristic search approach to planning with temporally extended preferences. *Artificial Intelligence*, 173(5–6):593–618, 2009. DOI: 10.1016/j.artint.2008.11.011 63

C. L. Baker, R. Saxe, and J. B. Tenenbaum. Action understanding as inverse planning. *Cognition*, 113(3):329–349, 2009. DOI: 10.1016/j.cognition.2009.07.005 59

R. K. Balla and A. Fern. UCT for tactical assault planning in real-time strategy games. In *Proc. 21st Int. Joint Conf. on Artificial Intelligence*, pages 40–45, 2009. 93

D. Ballard, M. Hayhoe, P. Pook, and R. Rao. Deictic codes for the embodiment of cognition. *Behavioral and Brain Sciences*, 20(4):723–742, 1997. DOI: 10.1017/S0140525X97001611 10, 61

A. Barto, S. Bradtke, and S. Singh. Learning to act using real-time dynamic programming. *Artificial Intelligence*, 72:81–138, 1995. DOI: 10.1016/0004-3702(94)00011-O 72, 86, 90, 104

A. Bauer and P. Haslum. LTL goal specifications revisited. In *Proc. 19th European Conf. on Artificial Intelligence*, pages 881–886, 2010. DOI: 10.3233/978-1-60750-606-5-881 63

R. Bellman. *Dynamic Programming*. Princeton University Press, 1957. 70, 81

D. S. Bernstein, R. Givan, N. Immerman, and S. Zilberstein. The complexity of decentralized control of markov decision processes. *Mathematics of Operations Research*, 27(4):819–840, 2002. DOI: 10.1287/moor.27.4.819.297 108, 109

P. Bertoli and A. Cimatti. Improving heuristics for planning as search in belief space. In *Proc. 6th Int. Conf. on Artificial Intelligence Planning Systems*, pages 143–152, 2002. 71

P. Bertoli, A. Cimatti, M. Roveri, and P. Traverso. Planning in nondeterministic domains under partial observability via symbolic model checking. In *Proc. 17th Int. Joint Conf. on Artificial Intelligence*, pages 473–478, 2001. 73

P. Bertoli, A. Cimatti, M. Pistore, and P. Traverso. A framework for planning with extended goals under partial observability. In *Proc. 13th Int. Conf. on Automated Planning and Scheduling*, pages 215–225, 2003. 62

D. P. Bertsekas and J. N. Tsitsiklis. *Parallel and Distributed Computation: Numerical Methods*. Prentice Hall, 1989. 85

D. P. Bertsekas and J. N. Tsitsiklis. *Neuro-Dynamic Programming*. Athena Scientific, 1996. 96, 104

D. P. Bertsekas. *Dynamic Programming and Optimal Control, Vols 1 and 2*. Athena Scientific, 1995. 6, 31, 70, 76, 79, 81, 82, 84, 85

A. Biere, M. Heule, H. Van Maaren, and T. Walsh, editors. *Handbook of Satisfiability: Frontiers in Artificial Intelligence and Applications*. IOS Press, 2012. 45, 110

A. Blum and M. Furst. Fast planning through planning graph analysis. In *Proc. 14th Int. Joint Conf. on Artificial Intelligence*, pages 1636–1642, 1995. DOI: 10.1016/S0004-3702(96)00047-1 13, 31, 45, 49

B. Bonet and H. Geffner. Planning as heuristic search: New results. In *Proc. 5th European Conf. on Planning*, pages 359–371, 1999. DOI: 10.1007/10720246_28 44, 45

B. Bonet and H. Geffner. Planning with incomplete information as heuristic search in belief space. In *Proc. 5th Int. Conf. on Artificial Intelligence Planning Systems*, pages 52–61, 2000. 55, 71, 77

B. Bonet and H. Geffner. Planning as heuristic search. *Artificial Intelligence*, 129(1–2):5–33, 2001. DOI: 10.1016/S0004-3702(01)00108-4 27, 30, 31, 41

B. Bonet and H. Geffner. Faster heuristic search algorithms for planning with uncertainty and full feedback. In *Proc. 18th Int. Joint Conf. on Artificial Intelligence*, pages 1233–1238, 2003. 88, 89, 91

B. Bonet and H. Geffner. Labeled RTDP: Improving the convergence of real-time dynamic programming. In *Proc. 13th Int. Conf. on Automated Planning and Scheduling*, pages 12–31, 2003. 91

B. Bonet and H. Geffner. mGPT: A probabilistic planner based on heuristic search. *Journal of Artificial Intelligence Research*, 24:933–944, 2005. DOI: 10.1613/jair.1688 77

B. Bonet and H. Geffner. Solving POMDPs: RTDP-Bel vs. point-based algorithms. In *Proc. 21st Int. Joint Conf. on Artificial Intelligence*, pages 1641–1646, 2009. 83, 99, 104, 105

B. Bonet and H. Geffner. Planning under partial observability by classical replanning: Theory and experiments. In *Proc. 22nd Int. Joint Conf. on Artificial Intelligence*, pages 1936–1941, 2011. DOI: 10.5591/978-1-57735-516-8/IJCAI11-324 57, 73

B. Bonet and H. Geffner. Action selection for MDPs: Anytime AO* versus UCT. In *Proc. 26nd Conf. on Artificial Intelligence*, pages 1749–1755, 2012. 73, 94

B. Bonet and H. Geffner. Width and complexity of belief tracking in non-deterministic conformant and contingent planning. In *Proc. 26nd Conf. on Artificial Intelligence*, pages 1756–1762, 2012. 74

B. Bonet and H. Geffner. Causal belief decomposition for planning with sensing: Completeness and practical approximation. In *Proc. 23rd Int. Joint Conf. on Artificial Intelligence*, 2013. 75, 76

B. Bonet and M. Helmert. Strengthening landmark heuristics via hitting sets. In *Proc. 19th European Conf. on Artificial Intelligence*, pages 329–334, 2010. 42

B. Bonet, G. Loerincs, and H. Geffner. A robust and fast action selection mechanism for planning. In *Proc. 14th Nat. Conf. on Artificial Intelligence*, pages 714–719, 1997. 13, 24, 30

B. Bonet, H. Palacios, and H. Geffner. Automatic derivation of memoryless policies and finite-state controllers using classical planners. In *Proc. 19th Int. Conf. on Automated Planning and Scheduling*, pages 34–41, 2009. 10, 60, 61, 62

B. Bonet. Conformant plans and beyond: Principles and complexity. *Artificial Intelligence*, 174:245–269, 2010. DOI: 10.1016/j.artint.2009.11.001 54

M. Botvinick and J. An. Goal-directed decision making in the prefrontal cortex: a computational framework. In *Proc. 22nd Annual Conf. on Advances in Neural Information Processing Systems*, pages 169–176, 2008. 108

C. Boutilier, T. Dean, and S. Hanks. Decision-theoretic planning: Structural assumptions and computational leverage. *Journal of Artificial Intelligence Research*, 1:1–93, 1999. DOI: 10.1613/jair.575 79

C. Boutilier, R. Reiter, and B. Price. Symbolic dynamic programming for first-order MDPs. In *Proc. 17th Int. Joint Conf. on Artificial Intelligence*, pages 690–700, 2001. 108

M. Bowling, R. Jensen, and M. Veloso. A formalization of equilibria for multiagent planning. In *Proc. 18th Int. Joint Conf. on Artificial Intelligence*, pages 1460–1462, 2003. 109

R. I. Brafman and C. Domshlak. Factored planning: How, when, and when not. In *Proc. 21st Nat. Conf. on Artificial Intelligence*, pages 809–814, 2006. 35

R. I. Brafman and G. Shani. A multi-path compilation approach to contingent planning. In *Proc. 26nd Conf. on Artificial Intelligence*, pages 1868–1874, 2012. 57

R. I. Brafman and G. Shani. Replanning in domains with partial information and sensing actions. *Journal of Artificial Intelligence Research*, 1(45):565–600, 2012. DOI: 10.1613/jair.3711 55, 57, 73

R. I. Brafman and M. Tennenholtz. R-max-a general polynomial time algorithm for near-optimal reinforcement learning. *Journal of Machine Learning Research*, 3:213–231, 2003. DOI: 10.1162/153244303765208377 96

R. I. Brafman, C. Domshlak, Y. Engel, and M. Tennenholtz. Planning games. In *Proc. 21st Int. Joint Conf. on Artificial Intelligence*, pages 73–78, 2009. 109

D. Bryce, S. Kambhampati, and D. E. Smith. Planning graph heuristics for belief space search. *Journal of Artificial Intelligence Research*, 26:35–99, 2006. DOI: 10.1613/jair.1869 55, 71, 73

M. Buckland. *Programming Game AI by Example*. Wordware Publishing, Inc., 2004. 10

T. Bylander. The computational complexity of propositional STRIPS planning. *Artificial Intelligence*, 69:165–204, 1994. DOI: 10.1016/0004-3702(94)90081-7 8, 16, 34, 35

P. P. Chakrabarti, S. Ghose, and S. C. De Sarkar. Best first search in AND/OR graphs. In *Proc. 16th Annual ACM Conf. on Computer Science*, pages 256–261, 1988. DOI: 10.1145/322609.322650 94

D. Chapman. Penguins can make cake. *AI Magazine*, 10(4):45–50, 1989. 10, 12, 61

G. M. J. Chaslot, M. H. M. Winands, H. Herik, J. Uiterwijk, and B. Bouzy. Progressive strategies for Monte-Carlo tree search. *New Mathematics and Natural Computation*, 4(3):343–357, 2008. DOI: 10.1142/S1793005708001094 93

H. Chen and O. Giménez. Act local, think global: Width notions for tractable planning. In *Proc. 17th Int. Conf. on Automated Planning and Scheduling*, pages 73–80, 2007. 35

A. Cimatti, M. Pistore, M. Roveri, and P. Traverso. Weak, strong, and strong cyclic planning via symbolic model checking. *Artificial Intelligence*, 147(1):35–84, 2003. DOI: 10.1016/S0004-3702(02)00374-0 76, 108

A. Cimatti, M. Roveri, and P. Bertoli. Conformant planning via symbolic model checking and heuristic search. *Artificial Intelligence*, 159:127–206, 2004. DOI: 10.1016/j.artint.2004.05.003 55, 74

E. M. Clarke, O. Grumberg, and D. A. Peled. *Model Checking*. MIT Press, 2000. 108

A. J. Coles, A. Coles, M. Fox, and D. Long. Temporal planning in domains with linear processes. In *Proc. 21st Int. Joint Conf. on Artificial Intelligence*, pages 1671–1676, 2009. 48

A. J. Coles, A. Coles, A. García Olaya, S. Jiménez, C. Linares López, S. Sanner, and S. Yoon. A survey of the seventh international planning competition. *AI Magazine*, 33(1):83–88, 2012. 34, 39

T. Cormen, C. Leiserson, R. Rivest, and C. Stein. *Introduction to Algorithms*. MIT Press, 3rd edition, 2009. 6, 16, 17, 31

S. Cresswell and A. M. Coddington. Compilation of LTL goal formulas into PDDL. In *Proc. 16th European Conf. on Artificial Intelligence*, pages 985–986, 2004. 63

J. Culberson and J. Schaeffer. Pattern databases. *Computational Intelligence*, 14(3):318–334, 1998. DOI: 10.1111/0824-7935.00065 41

W. Cushing, S. Kambhampati, Mausam, and D. S. Weld. When is temporal planning really temporal? In *Proc. 20th Int. Joint Conf. on Artificial Intelligence*, pages 1852–1859, 2007. 49

M. Daniele, P. Traverso, and M. Y. Vardi. Strong cyclic planning revisited. In *Proc. 5th European Conf. on Planning*, pages 35–48, 1999. DOI: 10.1007/10720246_3 65, 76, 77

G. de Giacomo and M. Y. Vardi. Automata-theoretic approach to planning for temporally extended goals. In *Proc. 5th European Conf. on Planning*, pages 226–238, 1999. DOI: 10.1007/10720246_18 62

T. Dean, L. P. Kaelbling, J. Kirman, and A. Nicholson. Planning with deadlines in stochastic domains. In *Proc. 11th Nat. Conf. on Artificial Intelligence*, pages 574–579, 1993. 13

R. Dechter, I. Meiri, and J. Pearl. Temporal constraint networks. *Artificial Intelligence*, 49:61–95, 1991. DOI: 10.1016/0004-3702(91)90006-6 47

R. Dechter. *Constraint Processing*. Morgan Kaufmann, 2003. 35, 46

D. C. Dennett. *Kinds of minds*. Basic Books New York, 1996. 2

E. Dijkstra. A note on two problems in connexion with graphs. *Numerische mathematik*, 1(1):269–271, 1959. DOI: 10.1007/BF01386390 6

M. B. Do and S. Kambhampati. Solving the planning-graph by compiling it into CSP. In *Proc. 5th Int. Conf. on Artificial Intelligence Planning Systems*, pages 82–91, 2000. 46

M. B. Do and S. Kambhampati. Sapa: A domain-independent heuristic metric temporal planner. In *Proc. 6th European Conf. on Planning*, pages 82–91, 2001. 30, 48

A. Doucet, N. de Freitas, K. Murphy, and S. Russell. Rao-blackwellised particle filtering for dynamic bayesian networks. In *Proc. 16th Conf. on Uncertainty on Artificial Intelligence*, pages 176–183, 2000. 107

S. Edelkamp and S. Schrödl. *Heuristic Search – Theory and Applications*. Academic Press, 2012. 17, 19, 23, 108

S. Edelkamp. Planning with pattern databases. In *Proc. 6th European Conf. on Planning*, 2001. 41

S. Edelkamp. On the compilation of plan constraints and preferences. In *Proc. 16th Int. Conf. on Automated Planning and Scheduling*, pages 374–377, 2006. 48, 63

K. Erol, J. Hendler, and D. S. Nau. HTN planning: Complexity and expressivity. In *Proc. 12th Nat. Conf. on Artificial Intelligence*, pages 1123–1123, 1994. 11, 49

P. Eyerich, T. Keller, and M. Helmert. High-quality policies for the canadian traveler's problem. In *Proc. 24th Conf. on Artificial Intelligence*, pages 51–58, 2010. 94

Z. Feng and E. A. Hansen. Symbolic heuristic search for factored Markov decision processes. In *Proc. 16th Nat. Conf. on Artificial Intelligence*, pages 455–460, 1999. 108

Z. Feng, E. A. Hansen, and S. Zilberstein. Symbolic generalization for on-line planning. In *Proc. 18th Conf. on Uncertainty on Artificial Intelligence*, pages 209–216, 2002. 108

A. Fern, S. Yoon, and R. Givan. Approximate policy iteration with a policy language bias. In *Proc. 17th Annual Conf. on Advances in Neural Information Processing Systems*, 2003. DOI: 10.1613/jair.1700 12, 110

R. Fikes and N. Nilsson. STRIPS: A new approach to the application of theorem proving to problem solving. *Artificial Intelligence*, 1:27–120, 1971. DOI: 10.1016/0004-3702(71)90010-5 12, 24

H. Finnsson and Y. Björnsson. Simulation-based approach to general game playing. In *Proc. 23th Conf. on Artificial Intelligence*, pages 259–264, 2008. 93

M. Fox and D. Long. PDDL 2.1: An extension to PDDL for expressing temporal planning domains. *Journal of Artificial Intelligence Research*, 20:61–124, 2003. 49

E. Freuder. A sufficient condition for backtrack-free search. *Journal of the ACM*, 29(1):24–32, 1982. DOI: 10.1145/322290.322292 35

J. Fu, V. Ng, F. Bastani, and I. Yen. Simple and fast strong cyclic planning for fully-observable nondeterministic planning problems. In *Proc. 22nd Int. Joint Conf. on Artificial Intelligence*, pages 1949–1954, 2011. DOI: 10.5591/978-1-57735-516-8/IJCAI11-326 78

B. Gazen and C. Knoblock. Combining the expressiveness of UCPOP with the efficiency of Graphplan. In *Proc. 4th European Conf. on Planning*, pages 221–233, 1997. 26, 51

H. Geffner and B. Bonet. Solving large POMDPs using real time dynamic programming, 1998. AAAI Fall Symposium on POMDPs. 104

H. Geffner. Heuristics, planning, cognition. In R. Dechter, H. Geffner, and J. Y. Halpern, editors, *Heuristics, Probability and Causality. A Tribute to Judea Pearl*. College Publications, 2010. 112

H. Geffner. Artificial Intelligence: From Programs to Solvers. *AI Communications*, 2013. DOI: 10.3233/978-1-58603-925-7-4 8

H. Geffner. Computational models of planning. *Wiley Interdisciplinary Reviews: Cognitive Science*, 2, 2013. DOI: 10.1002/wcs.1233 112

C. W. Geib and R. P. Goldman. A probabilistic plan recognition algorithm based on plan tree grammars. *Artificial Intelligence*, 173(11):1101–1132, 2009. DOI: 10.1016/j.artint.2009.01.003 57

S. Gelly and D. Silver. Combining online and offline knowledge in UCT. In *Proc. 24th Int. Conf. on Machine Learning*, pages 273–280, 2007. DOI: 10.1145/1273496.1273531 93

A. E. Gerevini, A. Saetti, and I. Serina. An approach to efficient planning with numerical fluents and multi-criteria plan quality. *Artificial Intelligence*, 172(8):899–944, 2008. DOI: 10.1016/j.artint.2008.01.002 48

A. E. Gerevini, P. Haslum, D. Long, A. Saetti, and Y. Dimopoulos. Deterministic planning in the fifth international planning competition: PDDL3 and experimental evaluation of the planners. *Artificial Intelligence*, 173(5–6):619–668, 2009. DOI: 10.1016/j.artint.2008.10.012 25, 62

R. Gerth, D. A. Peled, M. Y. Vardi, and P. Wolper. Simple on-the-fly automatic verification of linear temporal logic. In *Proc. Int. Symposium on Protocol Specification, Testing and Verification*, pages 3–18, 1995. 64

M. Ghallab, D. Nau, and P. Traverso. *Automated Planning: theory and practice*. Morgan Kaufmann, 2004. xi, 13, 46, 49

G. Gigerenzer. *Gut feelings: The intelligence of the unconscious*. Viking Books, 2007. 111

M. L. Ginsberg. Universal planning: An (almost) universally bad idea. *AI Magazine*, 10(4):40–44, 1989. 12

R. P. Goldman and M. S. Boddy. Expressive planning and explicit knowledge. In *Proc. 3rd Int. Conf. on Artificial Intelligence Planning Systems*, pages 110–117, 1996. 54

A. Gopnik, C. Glymour, D. Sobel, L. Schulz, T. Kushnir, and D. Danks. A theory of causal learning in children: Causal maps and bayes nets. *Psychological Review*, 111(1):3–31, 2004. DOI: 10.1037/0033-295X.111.1.3 111

E. A. Hansen and R. Zhou. Anytime heuristic search. *Journal of Artificial Intelligence Research*, 28:267–297, 2007. DOI: 10.1613/jair.2096 19

E. A. Hansen and S. Zilberstein. LAO*: A heuristic search algorithm that finds solutions with loops. *Artificial Intelligence*, 129:35–62, 2001. DOI: 10.1016/S0004-3702(01)00106-0 72, 91

E. A. Hansen. Solving POMDPs by searching in policy space. In *Proc. 14th Conf. on Uncertainty on Artificial Intelligence*, pages 211–219, 1998. 107

P. Hart, N. Nilsson, and B. Raphael. A formal basis for the heuristic determination of minimum cost paths. *IEEE Trans. on Systems Science and Cybernetics*, 4:100–107, 1968. DOI: 10.1109/TSSC.1968.300136 17

P. Haslum and H. Geffner. Admissible heuristics for optimal planning. In *Proc. 5th Int. Conf. on Artificial Intelligence Planning Systems*, pages 70–82, 2000. 13, 33, 41

P. Haslum and P. Jonsson. Some results on the complexity of planning with incomplete information. In *Proc. 5th European Conf. on Planning*, pages 308–318, 1999. DOI: 10.1007/10720246_24 54

P. Haslum, A. Botea, M. Helmert, B. Bonet, and S. Koenig. Domain-independent construction of pattern database heuristics for cost-optimal planning. In *Proc. 22nd Conf. on Artificial Intelligence*, pages 1007–1012, 2007. 42

R. Hassin, J. Uleman, and J. Bargh. *The New Unconscious*. Oxford University Press, 2005. 111

M. Helmert and C. Domshlak. Landmarks, critical paths and abstractions: What's the difference anyway? In *Proc. 19th Int. Conf. on Automated Planning and Scheduling*, pages 162–169, 2009. 41, 42

M. Helmert, P. Haslum, and J. Hoffmann. Flexible abstraction heuristics for optimal sequential planning. In *Proc. 17th Int. Conf. on Automated Planning and Scheduling*, pages 176–183, 2007. 42

M. Helmert, M. B. Do, and I. Refanidis. 2008 IPC Deterministic planning competition. In *6th Int. Planning Competition Booklet (ICAPS 2008)*, 2008. 39, 51

M. Helmert. The Fast Downward planning system. *Journal of Artificial Intelligence Research*, 26:191–246, 2006. DOI: 10.1613/jair.1705 33, 38, 39

M. Helmert. Concise finite-domain representations for PDDL planning tasks. *Artificial Intelligence*, 173(5):503–535, 2009. DOI: 10.1016/j.artint.2008.10.013 26

J. Hoey, R. St-Aubin, A. Hu, and C. Boutilier. SPUDD: Stochastic planning using decision diagrams. In *Proc. 15th Conf. on Uncertainty on Artificial Intelligence*, pages 279–288, 1999. 108

J. Hoffmann and R. I. Brafman. Contingent planning via heuristic forward search with implicit belief states. In *Proc. 15th Int. Conf. on Automated Planning and Scheduling*, pages 71–80, 2005. 72

J. Hoffmann and R. I. Brafman. Conformant planning via heuristic forward search: A new approach. *Artificial Intelligence*, 170:507–541, 2006. DOI: 10.1016/j.artint.2006.01.003 55, 73

J. Hoffmann and B. Nebel. The FF planning system: Fast plan generation through heuristic search. *Journal of Artificial Intelligence Research*, 14:253–302, 2001. DOI: 10.1613/jair.855 13, 31, 37, 39

J. Hoffmann, J. Porteous, and L. Sebastia. Ordered landmarks in planning. *Journal of Artificial Intelligence Research*, 22:215–278, 2004. DOI: 10.1613/jair.1492 13, 38, 39

J. Hoffmann, C. Gomes, B. Selman, and H. A. Kautz. SAT encodings of state-space reachability problems in numeric domains. In *Proc. 20th Int. Joint Conf. on Artificial Intelligence*, pages 1918–1923, 2007. 34, 46

J. Hoffmann. The Metric-FF planning system: Translating "ignoring delete lists" to numeric state variables. *Journal of Artificial Intelligence Research*, 20:291–341, 2003. DOI: 10.1613/jair.1144 48

J. Hoffmann. Where 'ignoring delete lists' works: Local search topology in planning benchmarks. *Journal of Artificial Intelligence Research*, 24:685–758, 2005. DOI: 10.1613/jair.1747 35

J. Hoffmann. Analyzing search topology without running any search: On the connection between causal graphs and h^+. *Journal of Artificial Intelligence Research*, 41:155–229, 2011. DOI: 10.1613/jair.3276 35

J. Hopcroft and J. Ullman. *Introduction to Automata Theory, Languages, and Computation*. Addison-Wesley, 1979. 64

R. Howard. *Dynamic Probabilistic Systems – Volume I: Markov Models*. Wiley, 1971. 85

Y. Hu and G. de Giacomo. Generalized planning: Synthesizing plans that work for multiple environments. In *Proc. 22nd Int. Joint Conf. on Artificial Intelligence*, pages 918–923, 2011. DOI: 10.5591/978-1-57735-516-8/IJCAI11-159 11

P. Jonsson and C. Bäckström. Tractable planning with state variables by exploiting structural restrictions. In *Proc. 12th Nat. Conf. on Artificial Intelligence*, pages 998–1003, 1994. 35

A. Jonsson. The role of macros in tractable planning over causal graphs. In *Proc. 20th Int. Joint Conf. on Artificial Intelligence*, pages 1936–1941, 2007. 111

A. Junghanns and J. Schaeffer. Sokoban: Enhancing general single-agent search methods using domain knowledge. *Artificial Intelligence*, 129(1):219–251, 2001. DOI: 10.1016/S0004-3702(01)00109-6 23

F. Kabanza and S. Thiébaux. Search control in planning for temporally extended goals. In *Proc. 15th Int. Conf. on Automated Planning and Scheduling*, pages 130–139, 2005. 64

L. P. Kaelbling, M. L. Littman, and A. Cassandra. Planning and acting in partially observable stochastic domains. *Artificial Intelligence*, 101(1–2):99–134, 1998. DOI: 10.1016/S0004-3702(98)00023-X 6, 13, 101, 107

Daniel Kahneman. *Thinking, fast and slow*. Farrar, Straus and Giroux, 2011. 111

S. Kambhampati, C. Knoblock, and Q. Yang. Planning as refinement search: A unified framework for evaluating design tradeoffs in partial-order planning. *Artificial Intelligence*, 76(1–2):167–238, 1995. DOI: 10.1016/0004-3702(94)00076-D 47

E. Karpas and C. Domshlak. Cost-optimal planning with landmarks. In *Proc. 21st Int. Joint Conf. on Artificial Intelligence*, pages 1728–1733, 2009. 41

M. Katz and C. Domshlak. Optimal additive composition of abstraction-based admissible heuristics. In *Proc. 18th Int. Conf. on Automated Planning and Scheduling*, pages 174–181, 2008. DOI: 10.1016/j.artint.2010.04.021 42

M. Katz and C. Domshlak. Structural patterns heuristics via fork decomposition. In *Proc. 18th Int. Conf. on Automated Planning and Scheduling*, pages 182–189, 2008. 42

H. A. Kautz and J. F. Allen. Generalized plan recognition. In *Proc. 5th Nat. Conf. on Artificial Intelligence*, pages 32–37, 1986. 57

H. A. Kautz and B. Selman. Planning as satisfiability. In *Proc. 10th European Conf. on Artificial Intelligence*, pages 359–363, 1992. 45

H. A. Kautz and B. Selman. Pushing the envelope: Planning, propositional logic, and stochastic search. In *Proc. 13th Nat. Conf. on Artificial Intelligence*, pages 1194–1201, 1996. 13, 45, 110

H. A. Kautz and B. Selman. Unifying SAT-based and graph-based planning. In *Proc. 16th Int. Joint Conf. on Artificial Intelligence*, pages 318–327, 1999. 45, 46

T. Keller and P. Eyerich. PROST: Probabilistic planning based on UCT. In *Proc. 22nd Int. Conf. on Automated Planning and Scheduling*, pages 119–127, 2012. 94

E. Keyder and H. Geffner. Heuristics for planning with action costs revisited. In *Proc. 18th European Conf. on Artificial Intelligence*, pages 588–592, 2008. DOI: 10.3233/978-1-58603-891-5-588 32

E. Keyder and H. Geffner. The HMDPP planner for planning with probabilities. In *6th Int. Planning Competition Booklet (ICAPS 2008)*, 2008. 93

E. Keyder and H. Geffner. Soft goals can be compiled away. *Journal of Artificial Intelligence Research*, 36:547–556, 2009. DOI: 10.1613/jair.2857 52, 53

E. Keyder, S. Richter, and M. Helmert. Sound and complete landmarks for And/Or graphs. In *Proc. 19th European Conf. on Artificial Intelligence*, pages 335–340, 2010. 39

C. A. Knoblock. Learning abstraction hierarchies for problem solving. In *Proc. 8th Nat. Conf. on Artificial Intelligence*, pages 923–928, 1990. 111

L. Kocsis and C. Szepesvári. Bandit based Monte-Carlo planning. In *Proc. 17th European Conf. on Machine Learning*, pages 282–293, 2006. DOI: 10.1007/11871842_29 93

S. Koenig and X. Sun. Comparing real-time and incremental heuristic search for real-time situated agents. *Journal of Autonomous Agents and Multi-Agent Systems*, 18(3):313–341, 2009. DOI: 10.1007/s10458-008-9061-x 23

A. Kolobov, P. Dai, Mausam, and D. S. Weld. Reverse iterative deepening for finite-horizon MDPs with large branching factors. In *Proc. 22nd Int. Conf. on Automated Planning and Scheduling*, pages 146–154, 2012. 93

A. Kolobov, Mausam, and D. S. Weld. LRTDP versus UCT for online probabilistic planning. In *Proc. 26nd Conf. on Artificial Intelligence*, pages 1786–1792, 2012. 93

R. E. Korf. Depth-first iterative-deepening: An optimal admissible tree search. *Artificial Intelligence*, 27(1):97–109, 1985. DOI: 10.1016/0004-3702(85)90084-0 19

R. E. Korf. Planning as search: A quantitative approach. *Artificial Intelligence*, 33(1):65–88, 1987. DOI: 10.1016/0004-3702(87)90051-8 34, 111

R. E. Korf. Real-time heuristic search. *Artificial Intelligence*, 42:189–211, 1990. DOI: 10.1016/0004-3702(90)90054-4 21, 89

U. Kuter, D. S. Nau, E. Reisner, and R. P. Goldman. Using classical planners to solve nondeterministic planning problems. In *Proc. 18th Int. Conf. on Automated Planning and Scheduling*, pages 190–197, 2008. 78

E. L. Lawler, J. K. Lenstra, A. H. G. Rinnooy Kan, and D. B. Shmoys, editors. *The Traveling Salesman Problem: A Guided Tour of Combinatorial Optimization*. Wiley, 1985. 23, 43

N. Lipovetzky and H. Geffner. Searching for plans with carefully designed probes. In *Proc. 21st Int. Conf. on Automated Planning and Scheduling*, pages 154–161, 2011. 39

N. Lipovetzky and H. Geffner. Width and serialization of classical planning problems. In *Proc. 20th European Conf. on Artificial Intelligence*, pages 540–545, 2012. DOI: 10.3233/978-1-61499-098-7-540 35, 39

M. L. Littman, J. Goldsmith, and M. Mundhenk. The computational complexity of probabilistic planning. *Journal of Artificial Intelligence Research*, 9:1–36, 1998. DOI: 10.1613/jair.505 8

M. L. Littman. Memoryless policies: Theoretical limitations and practical results. In D. Cliff, editor, *From Animals to Animats 3*. MIT Press, 1994. 60

Y. Liu, S. Koenig, and D. Furcy. Speeding up the calculation of heuristics for heuristic search-based planning. In *Proc. 18th Nat. Conf. on Artificial Intelligence*, pages 484–491, 2002. 31

B. Marthi, S. J. Russell, and J. Wolfe. Angelic semantics for high-level actions. In *Proc. 17th Int. Conf. on Automated Planning and Scheduling*, pages 232–239, 2007. 111

M. Martin and H. Geffner. Learning generalized policies in planning using concept languages. In *Proc. 7th Int. Conf. on Principles of Knowledge Representation and Reasoning*, pages 667–677, 2000. 12

M. J. Mataric. *The Robotics Primer*. MIT Press, 2007. 10

Mausam and A. Kolobov. *Planning with Markov Decision Processes: An AI Perspective*. Morgan & Claypool, 2012. DOI: 10.2200/S00426ED1V01Y201206AIM017 92, 93

D. McAllester and D. Rosenblitt. Systematic nonlinear planning. In *Proc. 9th Nat. Conf. on Artificial Intelligence*, pages 634–639, 1991. 13, 47

D. V. McDermott, M. Ghallab, A. Howe, C. Knoblock, A. Ram, M. Veloso, D. S. Weld, and D. Wilkins. PDDL – The Planning Domain Definition Language. Technical Report CVC TR-98-003/DCS TR-1165, Yale Center for Computational Vision and Control, New Haven, CT, 1998. 26

D. V. McDermott. A heuristic estimator for means-ends analysis in planning. In *Proc. 3rd Int. Conf. on Artificial Intelligence Planning Systems*, pages 142–149, 1996. 13, 24

D. V. McDermott. Using regression-match graphs to control search in planning. *Artificial Intelligence*, 109(1–2):111–159, 1999. DOI: 10.1016/S0004-3702(99)00010-7 27, 31

124 BIBLIOGRAPHY

S. A. McIlraith and R. Fadel. Planning with complex actions. In *Proc. 9th Int. Workshop on Non-Monotonic Reasoning*, pages 356–364, 2002. 111

M. Minsky. Steps toward artificial intelligence. *Proceedings of the IRE*, 49(1):8–30, 1961. DOI: 10.1109/JRPROC.1961.287775 23

C. Muise, S. A. McIlraith, and J. C. Beck. Improved non-deterministic planning by exploiting state relevance. In *Proc. 22nd Int. Conf. on Automated Planning and Scheduling*, pages 172–180, 2012. 78

R. R. Murphy. *An Introduction to AI Robotics*. MIT Press, 2000. 10

B. Nebel. On the compilability and expressive power of propositional planning formalisms. *Journal of Artificial Intelligence Research*, 12:271–315, 2000. DOI: 10.1613/jair.735 26

A. Newell and H. A. Simon. GPS, a program that simulates human thought. In H. Billing, editor, *Lernende Automaten*, pages 109–124. R. Oldenbourg, 1961. 12

A. Newell, J. C. Shaw, and H. A. Simon. Report on a general problem-solving program. In *Proc. of the Int. Conf. on Information Processing*, pages 256–264, 1959. xi, 12

X. L. Nguyen and S. Kambhampati. Reviving partial order planning. In *Proc. 17th Int. Joint Conf. on Artificial Intelligence*, pages 459–466, 2001. 13, 47

N. Nilsson. *Principles of Artificial Intelligence*. Tioga, 1980. 44, 71, 91

H. Palacios and H. Geffner. Compiling uncertainty away in conformant planning problems with bounded width. *Journal of Artificial Intelligence Research*, 35:623–675, 2009. DOI: 10.1613/jair.2708 54, 55, 57, 73, 74

F. Patrizi, N. Lipovetzky, G. de Giacomo, and H. Geffner. Computing infinite plans for LTL goals using a classical planner. In *Proc. 22nd Int. Joint Conf. on Artificial Intelligence*, pages 2003–2008, 2011. DOI: 10.5591/978-1-57735-516-8/IJCAI11-334 64

F. Patrizi, N. Lipovetzky, and H. Geffner. Fair LTL synthesis for non-deterministic systems using strong cyclic planners. In *Proc. 23rd Int. Joint Conf. on Artificial Intelligence*, 2013. 64

J. Pearl. *Heuristics*. Addison-Wesley, 1983. 17, 18, 19, 23, 71

J. Pearl. *Probabilistic Reasoning in Intelligent Systems*. Morgan Kaufmann, 1988. 58, 74, 107

E. Pednault. ADL: Exploring the middle ground between Strips and the situation calculus. In *Proc. 1st Int. Conf. on Principles of Knowledge Representation and Reasoning*, pages 324–332, 1989. 26

J. Penberthy and D. S. Weld. UCPOP: A sound, complete, partiall order planner for ADL. In *Proc. 3rd Int. Conf. on Principles of Knowledge Representation and Reasoning*, pages 103–114, 1992. 12

J. Pineau, G. J. Gordon, and S. Thrun. Anytime point-based approximations for large POMDPs. *Journal of Artificial Intelligence Research*, 27:335–380, 2006. DOI: 10.1613/jair.2078 102

A. Pnueli. The temporal logic of programs. In *Proc. 18th Annual Symposium on the Foundations of Computer Science*, pages 46–57, 1977. DOI: 10.1109/SFCS.1977.32 62

P. Poupart and C. Boutilier. Bounded finite state controllers. In *Proc. 17th Annual Conf. on Advances in Neural Information Processing Systems*, pages 823–830, 2003. 107

M. Puterman. *Markov Decision Processes – Discrete Stochastic Dynamic Programming*. John Wiley and Sons, 1994. DOI: 10.1002/9780470316887 79, 82, 85

M. Ramírez and H. Geffner. Probabilistic plan recognition using off-the-shelf classical planners. In *Proc. 24th Conf. on Artificial Intelligence*, pages 1121–1126, 2010. 58, 59

M. Ramírez and H. Geffner. Goal recognition over POMDPs: Inferring the intention of a POMDP agent. In *Proc. 22nd Int. Joint Conf. on Artificial Intelligence*, pages 2009–2014, 2011. DOI: 10.5591/978-1-57735-516-8/IJCAI11-335 59

S. Richter and M. Westphal. The LAMA planner: Guiding cost-based anytime planning with landmarks. *Journal of Artificial Intelligence Research*, 39:127–177, 2010. DOI: 10.1613/jair.2972 13, 20, 33, 38, 39

S. Richter, M. Helmert, and M. Westphal. Landmarks revisited. In *Proc. 23th Conf. on Artificial Intelligence*, pages 975–982, 2008. 39

S. Richter, J. T. Thayer, and W. Ruml. The joy of forgetting: Faster anytime search via restarting. In *Proc. 20th Int. Conf. on Automated Planning and Scheduling*, pages 137–144, 2010. 20

J. Rintanen. Complexity of planning with partial observability. In *Proc. 14th Int. Conf. on Automated Planning and Scheduling*, pages 345–354, 2004. 54

J. Rintanen. Distance estimates for planning in the discrete belief space. In *Proc. 19th Nat. Conf. on Artificial Intelligence*, pages 525–530, 2004. 55

J. Rintanen. Planning as satisfiability: Heuristics. *Artificial Intelligence*, 193:45–86, 2012. DOI: 10.1016/j.artint.2012.08.001 46

G. Röger and M. Helmert. The more, the merrier: Combining heuristic estimators for satisficing planning. In *Proc. 20th Int. Conf. on Automated Planning and Scheduling*, pages 246–249, 2010. 38

S. J. Russell and P. Norvig. *Artificial Intelligence: A Modern Approach*. Prentice Hall, 3rd edition, 2009. xi, 1, 13, 107

E. D. Sacerdoti. Planning in a hierarchy of abstraction spaces. *Artificial Intelligence*, 5(2):115–135, 1974. DOI: 10.1016/0004-3702(74)90026-5 111

E. Sacerdoti. The nonlinear nature of plans. In *Proc. 4th Int. Joint Conf. on Artificial Intelligence*, pages 206–214, 1975. 13

R. Sanchez and S. Kambhampati. Planning graph heuristics for selecting objectives in oversubscription planning problems. In *Proc. 15th Int. Conf. on Automated Planning and Scheduling*, pages 192–201, 2005. 51

S. Sanner and C. Boutilier. Practical solution techniques for first-order MDPs. *Artificial Intelligence*, 173(5–6):748–788, 2009. DOI: 10.1016/j.artint.2008.11.003 108

S. Sanner and K. Kersting. Symbolic dynamic programming for first-order POMDPs. In *Proc. 24th Conf. on Artificial Intelligence*, pages 1140–1146, 2010. 108

M. J. Schoppers. Universal plans for reactive robots in unpredictable environments. In *Proc. 10th Int. Joint Conf. on Artificial Intelligence*, pages 1039–1046, 1987. 12

W. Schultz, P. Dayan, and P. R. Montague. A neural substrate of prediction and reward. *Science*, 275(5306):1593–1599, 1997. DOI: 10.1126/science.275.5306.1593 111

G. Shani, J. Pineau, and R. Kaplow. A survey of point-based POMDP solvers. *Journal of Autonomous Agents and Multi-Agent Systems*, pages 1–51, 2012. Online-First Article. DOI: 10.1007/s10458-012-9200-2 102, 103

D. Silver and J. Veness. Monte-Carlo planning in large POMDPs. In *Proc. 24th Annual Conf. on Advances in Neural Information Processing Systems*, pages 2164–2172, 2010. 105

H. A. Simon. A behavioral model of rational choice. *The Quarterly Journal of Economics*, 69(1):99–118, 1955. DOI: 10.2307/1884852 23

H. A. Simon. *The sciences of the artificial*. MIT Press, 3rd edition, 1996. 34

M. Sipser. *Introduction to Theory of Computation*. Thomson Course Technology, Boston, MA, 2nd edition, 2006. 8, 45, 64

R. Smallwood and E. Sondik. The optimal control of partially observable Markov processes over a finite horizon. *Operations Research*, 21:1071–1088, 1973. DOI: 10.1287/opre.21.5.1071 98

D. E. Smith and D. S. Weld. Conformant graphplan. In *Proc. 15th Nat. Conf. on Artificial Intelligence*, pages 889–896, 1998. 54

D. E. Smith and D. S. Weld. Temporal planning with mutual exclusion reasoning. In *Proc. 16th Int. Joint Conf. on Artificial Intelligence*, pages 326–337, 1999. 49

D. E. Smith, J. Frank, and A. K. Jonsson. Bridging the gap between planning and scheduling. *The Knowledge Engineering Review*, 15(1):47–83, 2000. DOI: 10.1017/S0269888900001089 47, 48

D. E. Smith. Choosing objectives in over-subscription planning. In *Proc. 14th Int. Conf. on Automated Planning and Scheduling*, pages 393–401, 2004. 30, 51

S. Srivastava, N. Immerman, and S. Zilberstein. A new representation and associated algorithms for generalized planning. *Artificial Intelligence*, 175(2):615–647, 2011. DOI: 10.1016/j.artint.2010.10.006 11

S. Srivastava, S. Zilberstein, N. Immerman, and H. Geffner. Qualitative numeric planning. In *Proc. 25th Conf. on Artificial Intelligence*, pages 1010–1016, 2011. 111

K. Stanley, B. Bryant, and R. Miikkulainen. Real-time neuroevolution in the NERO video game. *IEEE Trans. on Evolutionary Computation*, 9(6):653–668, 2005. DOI: 10.1109/TEVC.2005.856210 12

C. Striebel. Sufficient statistics in the control of stochastic systems. *Journal of Mathematical Analaysis and Applications*, 12:576–592, 1965. DOI: 10.1016/0022-247X(65)90027-2 98

R. Sutton and A. Barto. *Introduction to Reinforcement Learning*. MIT Press, 1998. 95, 96, 104

C. Szepesvári. *Algorithms for reinforcement learning*. Morgan & Claypool Publishers, 2010. DOI: 10.2200/S00268ED1V01Y201005AIM009 96

R. E. Tarjan. Depth first search and linear graph algorithms. *SIAM Journal on Computing*, 1(2):146–160, 1972. DOI: 10.1137/0201010 91

A. Tate. Generating project networks. In *Proc. 5th Int. Joint Conf. on Artificial Intelligence*, pages 888–893, 1977. 13

S. T. To, E. Pontelli, and T. Cao Son. On the effectiveness of CNF and DNF representations in contingent planning. In *Proc. 22nd Int. Joint Conf. on Artificial Intelligence*, pages 2033–2038, 2011. DOI: 10.5591/978-1-57735-516-8/IJCAI11-339 55, 73

J. Tooby and L. Cosmides. The psychological foundations of culture. In J. Barkow, L. Cosmides, and J. Tooby, editors, *The Adapted Mind*. Oxford, 1992. 8

M. Toussaint and A. Storkey. Probabilistic inference for solving discrete and continuous state markov decision processes. In *Proc. 23rd Int. Conf. on Machine Learning*, pages 945–952, 2006. DOI: 10.1145/1143844.1143963 108

M. van den Briel and S. Kambhampati. Optiplan: Unifying IP-based and graph-based planning. *Journal of Artificial Intelligence Research*, 24:919–931, 2005. DOI: 10.1613/jair.1698 48

M. Y. Vardi and P. Wolper. Reasoning about infinite computations. *Information and Computation*, 115(1):1–37, 1994. DOI: 10.1006/inco.1994.1092 64

J. Veness, K. S. Ng, M. Hutter, W. Uther, and D. Silver. A Monte-Carlo AIXI approximation. *Journal of Artificial Intelligence Research*, 40:95–142, 2011. DOI: 10.1613/jair.3125 111

V. Vidal and H. Geffner. Branching and pruning: An optimal temporal POCL planner based on constraint programming. *Artificial Intelligence*, 170(3):298–335, 2006. DOI: 10.1016/j.artint.2005.08.004 13, 47, 48

C. Watkins. *Learning from Delayed Rewards*. PhD thesis, Cambridge University, 1989. 95

D. S. Weld. An introduction to least commitment planning. *AI Magazine*, 15(4):27–61, 1994. 44, 46, 47

T. D. Wilson. *Strangers to Ourselves*. Belknap Press, 2002. 111

Q. Yang. Activity recognition: Linking low-level sensors to high-level intelligence. In *Proc. 21st Int. Joint Conf. on Artificial Intelligence*, pages 20–25, 2009. 57

S. Yoon, A. Fern, and R. Givan. FF-replan: A baseline for probabilistic planning. In *Proc. 17th Int. Conf. on Automated Planning and Scheduling*, pages 352–359, 2007. 77, 93

L. Zhu and R. Givan. Landmark extraction via planning graph propagation. In *ICAPS Doctoral Consortium*, pages 156–160, 2003. 39

Authors' Biography

HECTOR GEFFNER

Hector is interested in artificial intelligence and cognitive science, having worked on both planning and plan recognition methods for generating and recognizing autonomous behavior using model-based methods. He is an ICREA Research Professor at the Universitat Pompeu Fabra in Barcelona where he heads the AI group and directs the Master in Intelligent Interactive Systems. He was born and grew up in Buenos Aires, and then obtained an EE degree from the Universidad Simón Bolívar in Caracas, and a Ph.D. in Computer Science from UCLA. He received the 1990 ACM Dissertation Award for a thesis done under the supervision of Judea Pearl, and the 2009 and 2010 ICAPS Influential Paper Awards. Hector is a fellow of AAAI and ECCAI, and Associate Editor of *Artificial Intelligence* and the *Journal of Artificial Intelligence Research*. He is the author of the book *Default Reasoning*, MIT Press, 1992, and co-editor with Rina Dechter and Joseph Halpern of the book *Heuristics, Probability and Causality: A Tribute to Judea Pearl*, College Publications, 2010.

BLAI BONET

Blai is a Professor in the Computer Science Department at Universidad Simón Bolívar in Caracas, Venezuela. His main research interests are automated planning, knowledge representation and search. He obtained B.Sc. and M.Sc. degrees in Computer Science from Universidad Simón Bolívar and a Ph.D. in Computer Science from UCLA. He received the 2009 ICAPS Influential Paper Award. Blai is Associate Editor of *Artificial Intelligence* and the *Journal of Artificial Intelligence Research*, and member of the Executive Council of ICAPS (Int. Conf. on Automated Planning and Scheduling).

Printed in the United States
by Baker & Taylor Publisher Services